We and the Universe:
astronomy
By Jonathan Cardoso

Thanks

I want to thank all those people who have always trusted me, believed in my work, and uphold me to get here. This work is a milestone in my life: it is my first book, which has been in the hands of many editors, these have done me the favor of not publishing it, in the same way I thank the person who stole me; today he is much better. I thank all my staff, my friends, and my readers.

Presentation

It is common to look at the sky at night, and to be fascine with the immensity of the stars that cover it, we are asked questions: where do we come from? Where are we going? What place do we occupy in the immensity of the Universe? What is the real nature of the Cosmos? What are these bright spots, these eyes peering at night? It is inhuman in the human being the desire to search, the desire to always want to know more, to always go beyond. The man in his walk on the timeline was able to work out a series of answers to these questions and always answer them; thus, creating other questions or beautiful stories. We know that astronomy is one of the oldest sciences because from the earliest times man learned to look up to the sky and tried to understand what he saw. The sun was already believed to be the center of the universe, or that comets were signs of evil omen, just as the winds, tides, rays, and all other phenomena of nature were already considered divine caprice.

Since 2009 I have been researching this subject for today, with great joy I present my first astronomy book, which comes to show that we all writers, scientists, philosophers, professional and amateur astronomers, homemakers, gardeners can discuss issues of the type why we and the Universe exist. A certainty: you, dear reader, will discover because of the dust we have come and the dust we will return to after all, we are all made of dust from the stars.

A Short History of Astronomy

- The first steps of astronomy -

The word used to designate astronomy is the junction of two Greek terms " astron ", astro and " nomos ", law. Its origin is confused with the origin of civilization itself, making it one of the oldest and most beautiful sciences; and there is no one who can contradict it. It could be said that there were several factors that contributed to the beginning of it, perhaps the most important was the need to cultivate to feed the people who were starting to settle and the creation of a calendar. It is also impossible to determine precisely when astrology was born, and it is also impossible to determine who was born first, astronomy or astrology, what is known is that the association of celestial events and daily life created in the minds of men a real importance the observation of the stars. Pseudoscience was widely used among the Egyptians; the Greeks were responsible for developing it; as well as the Roman Empire. In all ancient civilizations, it is possible to find certain astronomical knowledge. It is very likely that the first astronomical observations began with the phenomena of apparent rotation of the stars, from east to west. First, they observed that some stars are born to the east and set to the west, thus describing arcs of circumference above the horizon. Soon after, they noticed that some stars never even set, describing complete circumferences. Long observations led us to believe that the stars were fixed in a sphere around our planet. This sphere had as reference the axis of rotation that passes through the Earth, the points where this axis cuts this celestial sphere are the two poles north and south, of the Earth and the celestial sphere respectively. Knowledge of these poles was central to the orientation issue. Already in ancient

Egypt, Babylon, and China, near the third millennium BC, it was common knowledge that the alpha star of the constellation Draco barely moved at night because it was very close to the south pole. The discovery of the cardinal points, equator, parallel, and meridians was a consequence of the notion of the planetary axis. One of the greatest proofs that these notions were known (and used) before 2500 a. Are the pyramids of Egypt, which were built with their faces facing the cardinal points and some monuments of the old Babylon that were constructed with their edges pointed towards the same points. After the perceptions of the stellar movements, it was the turn of the Moon to be the celestial body to be studied. Today we know that the Moon takes about 27 days, 7 hours and 43 minutes to complete a sidereal revolution, which time the star uses to return to the same position in relation to the stars. To complete a synodic revolution, which is the time it takes to return to the same position relative to the Sun, the Moon employs about 29 days, 12 minutes, and 2.9 seconds. It is well known that since the ancient Babylonians these periods were already known, and quite accurately. After the discovery of the lunar periods, it was the turn of the Sun to have its studied movement. The perception of solar movements is very peculiar, because its movement is much less salient, because its movement is much slower and the most difficult is that the Sun cannot be seen at the same time as the other stars. For the study of this star, it was necessary to observe the stars that can be observed in the west and the rising sun. The ancient peoples observed that the stars observed shortly after the sun went down, were approaching the star-king, as time passed, until they disappeared; but they surged after the rising of the Sun. Thus, the ancient astronomers discovered the ecliptic, so they called the circumference between the stars where the Sun moved. The Egyptians knew well the apparent movement of the Sun and determined a fixed year, which was determined by the periodic return of the heliacal birth of the star Sothis (Alfa Canis Maroris, Sirius) when observed from the city of Heliopolis.

About four thousand years ago man learned to identify the stars and group them together with names of animals, objects, and gods: that is how the constellations came about. Among the constellations, the ancients observed the zodiac and these observations revealed the existence of certain stars that moved with a certain difference compared to other stars: they were the planets. The ancient Babylonians recognized this sinuous movement of the planets, as did the Egyptians.

- Greek astronomy -

It was around the sixth century BC that the Greeks appeared in the history of astronomy as a watershed: we do not know for sure the exact date, but it is known that it was they who gave a truly scientific imprint to development, religious and political order, in this way they believed that they could come to pure knowledge. Anaximander (611-547 BC) introduced in the astronomy the idea of the Earth's isolation in space. Pythagoras (6th century BC), along with Parmenides (514? - 450? AC) was the first thinker to postulate that our planet had the spherical shape. Around 340 BC Aristotle wrote a book which he called "On the Firmament" where he also postulated very interesting arguments to prove the sphericity of the Earth. Already Anaxagoras (c.500 - 482 BC) was the first to correctly interpret eclipses. The Greeks were the first to develop mathematical science and sought to frame the phenomena known in geometric schemes. Ideas postulated by the Pythagorean school, which lasted until the time of Kepler, said that the celestial movements should be circular and uniform. The planets were the objects of study that gave more work to the old thinkers, being m first place in the hall of astronomy like objects more studied until the middle of century XIX. For the thinker Plato, the major problem of astronomy was to represent and explain the movement that was observed on the planets.

Filolau (450-400 BC) who was a disciple of the Pythagorean school was the first to conceive of a system where the anomalies of planetary systems could be explained by a motion attributed to the Earth. This, should be considered as an ordinary planet that, along with nine other objects, revolved around a fire, which was situated in the center of the Universe. However, this system was very imperfect, and the idea of having the Earth out of the center was not very well accepted by other thinkers of the time.

Eudoxus de Anido (400 - 347 BC) elaborated a system that became known as Homocentric Sphere Theory. In this theory, the motion of each planet could be explained by the combined action of uniform rotations of various spheres. However, the system created by Eudoxus did not recognize the irregularities on the longitude of the Sun, which could be proved by the observation of the solstices. The observations only proved that the Homocentric Sphere Theory could not be sustained. 18 centuries before the Danish astronomer Tycho Brahe postulates the heliocentric system, it is probable that some astronomer of antiquity did it, although it did not last until our time there was no intact report that they had actually reached it, we have some reports that Heraclides of the Point (IV century BC) made use of the observations of the planets Mercury and Venus never move away from the Sun more than the limits already established, it determined that the center of the orbit of these planets could not be other than the Sun, of this form he created a partial heliocentric system because he observed that the brightness of the planet Mars was stronger when it was in opposition to the Sun, this could only be the indication that the center of the orbit of the same would be somewhere between the Earth and the Sun.

Aristarchus of Samos, according to the ocular testimony of Plutarch (c. 46-120) came to suppose that the sun was fixed and that the Earth revolving around him, but he could not explain mathematically his idea and it was too simple, not giving explain the movements of the Sun, the Moon, and the planets.

Who solved the problem was Aristotle, in his conception, the Earth was static and the sun and other stars were that moved around our planet. His conception was based on his mystical beliefs, because God would not go by another star but ours in the center of the Universe, already the movement should be the circular because it was the most perfect geometric form. Claudius Ptolemy formulated the idea of Aristotle in a more scientific and complete way. According to the Ptolemaic model, the Earth was at the center of the system and was surrounded by eight spheres (different from the 55 spheres of Aristotle's homocentric system). The first three spheres were diminished by the Moon, Mercury, and Venus; the other five were orbited by the Sun, Mars, Jupiter, Saturn, and the stars. Ptolemy considered the peculiarities of planetary orbits postulating that the planets would move in smaller orbits, around their respective spheres. The idea of the stars orbiting the last sphere was useful for this system because it was also useful for the church, since there was enough space after the last sphere to explain the Divine abode. This model stood out among the others because it could predict positions in the firmament with reasonable accuracy. Ptolemy also had flaws, and clear and grave flaws: the first was that, in order to accurately predict the positions of the celestial objects in the spheres, Ptolemy had to create a universal law, where this law affirmed that the Moon followed its orbit in a trajectory in which in some stages of its orbit, it would approach two times more of the Earth than in others. That is, from here on Earth we should see the Moon twice as big as in others. Ptolemy was the writer of the most important work of astronomy of the time, which was titled Syntaxe Mathematics (better known by the Arabic translation, "Almagest"). This book was important in many factors, but it also marks the decline of European astronomy and completely encloses Alexandrian astronomy, just as it marks the complete abandonment of astronomical observations.

"Divine omnipotence is manifested in the organization of the Universe." This quote is present in the Qur'an and because of its astronomy was raised to the highest of the sciences. After Muhammad created his vast empire, some caliphs created favorable conditions for the study of such a science. The Arabs were responsible for about six centuries of activities, six centuries where they built observatories, made improvements to existing instruments, also achieved superior accuracy in time measurement and simplified calculation methods. However, the Arabs did not create any theory and did not create any instrument, but they played a very important role in astronomy, because of their boards, most of the Greek knowledge came to us.

The Verified Tables, produced in the year 830 at the observatories of Damascus and Baghdad, are the result of several observations. As a result, they found that the ecliptic was not as oblique as thought (Thabit Ben Qurrah [908-946]), the solar apogee (al-Battomi [c. 858-929]), a star catalog, precious for magnitudes (al-Sufi [903-986]), an unevenness of the Moon's motion, perhaps the variation (Abulwefa [940-998].) According to CA Nallino, Venus's observations would have led to a representation of planetary movement equivalent to that of to admit an epicycle with a center in the Sun. In Cairo ibn Yunus (? -1009) and Alhazen (X-XI century) continued these works, and the first one built the Haquemite Tables which had a great reputation. In Spain, al-Zarqali (1029 And the first author of the Tables of Toledo, when Alfonso X (1223-1284) ordered the construction of the Tables Afonsinas made special appeal to Arab astronomers (c.1272). In Maragha, Nasir al-Din (1201-1274) which led to the construction of the Ilkanean Tables, and finally, in Samarkand, Prince Ulugh-Beg (1394-1449) left his name connected with astronomy, by a truly original stellar catalog.

- Nicolaus Copernicus -

There was a great need among the ancient thinkers to give a literal meaning to the Holy Scriptures, but this meant that they should give up all the knowledge the Greeks had taken years to acquire. In this way, all the works of Aristotle and Ptolemy were subjected to a rereading and the old theories began to be rethought.

It was a Polish monk named Mikolaj Kopernik (Nicholas Copernicus [Torun 1473 Frauenburg 1543]) who opened the doors to the revolution in astronomy. In the year 1491 he entered the University of Krakow to study drawing, mathematics, medicine and, under the guidance of Wojciech Brudzewski, studied astronomy. In the year 1496 Copernicus had to move to Italy, there spent ten years studying astronomy and in the year 1501 had to return to Poland. While passing through the Universities of Bologna, Padua and Ferrara he gathered all the knowledge of the time. There are reports that while he was in Bologna, he associated himself with Domenico Novarra (1454-1504) and it is believed that it was in his company that he made the first astronomical observations, one of which was the concealment of Aldebaran (Alfa Tauri) on March 9, 1497, and this was one of the few observations he made in his life. Copernicus spent many years studying celestial movements and by the year 1529, Nicholas published his book entitled Nic. Copernici de hipothesibus motuum coelestium a se constitutis comentariolus (Brief comments by Nicolaus Copernicus on his hypotheses about the celestial movements). In this work, Copernicus presented heliocentrism as an astronomical model, although he did not invent it. Although Pope Clement VII approved the teaching of theory in Rome and Cardinal Schönberg asked for its publication, Copernicus believed that he should still elaborate the theory a little better, it had to be superior to the Ptolemaic system, which had been standing for 14 centuries. The Polish monk not only feared the reprisal of the Catholic Church which was dominant at the time, as feared the criticism of other thinkers. It all started when a mathematics teacher from a locality called Rahetia went to Frauenburg to discuss Copernicus's ideas. Rätichus (this was his nickname) became a disciple of the monk and insisted that the Polish monk let his ideas be published. In the year 1540, Rätichus published the Narratio de libris revolutionum Copernici (Narrative on the works of Copernicus on the revolutions), and at the same time that he sent to Nürenberg the theory that Copernicus had worked for years, he adds. The book De revolutionibus orbium coelestium (The Revolutions of the Celestial World) became the first treatise on heliocentric astronomy, but Copernicus received the first copy of his work on his deathbed. Copernicus got what he wanted: with his work he made a revolution in the field of science and thought of the time, his work was so good that he managed to rival Ptolemy 's Almagest. In this work, the Polish thinker admits the double movement of the planets; around itself and around the Sun. One of the major problems of the Copernican system was that it postulated that the movement of the planets was circular and uniform. There is a great mystery today that involves the preface of the work: in his original work, Copernicus dedicated the book to Pope Paul III, and in this preface, he affirmed that nothing in that work was new and would only be a complement to facilitate the use of the Astronomical Tables, but the preface was changed and for many years the reputation of Copernicus and his theories were cast into doubt. That was when they discovered that it was changed by another, which was attributed to Andreas Osiander, who claimed to be only a hypothesis this system.

- Tycho Brahe -

The Danish astronomer Tycho Brahe (Knudstrup 1546 - Prague 1601) was the last astronomer with the naked eye and, after Hiparco, the most brilliant. It is believed that a solar eclipse announced for August 21, 1560 attracted his attention to the field of astronomy: at only 14 years of age, the forecast was fulfilled and he was amazed that man could predict the position of stars In the year 1563 he began to work on the Verification of the Pruducts Tables (1551) and discovered that they did not give the correct position of the Jupiter and Saturn planets. In the year 1572, Brahe observed the appearance of a new star in the constellation of Cassiopeia, which during the 18 months in which it was visible came to shine more than Venus and became visible even during the day. With the observation of this star, Brahe wrote a treaty that gave him a European reputation. In the book titled De nova stella, the Danish astronomer introduced the new vocabulary to designate stars that we now know to explode and become visible. In this work, the astronomer demonstrated that the new star was located farther away than the Moon, defeating the Aristotelian ideas on the inalterability and perfection of the celestial sphere.

Tycho Brahe was young and famous, and the king of Denmark Frederick II (1534 1588) ceded to him the island of Ven (former island of Hven) in the Sund. There they built the observatory of Uraninburg and this was the first astronomical observatory of history, equipped with the most accurate instruments of the time. In the year 1577, Brahe made strict observations of a great comet and demonstrated that this was not an atmospheric phenomenon, as had been believed since the time of Aristotle. determined the duration of the year with an error of less than one second, which caused the reform of the calendar in 1582 by Pope Gregory XIII; correcting the ten days of the Julian calendar (which had been in force since the Roman Empire). Johannes Kepler started working for Brahe. The main work of the Danish Astronomiae instauratae progymnasmata, was published by Kepler in 1602 and 1603 and was divided into two parts. The first part dealt with the movements of the Sun and the Moon, also contains a catalog of 777 fixed stars. The second part (De Mundi aetherii recentioribus Phoenomenis), this part had been completed in 1588 by Brahe, and dealt with the great comet that appeared in 1577. In this work he demonstrates that he found an insensitive parallax in the star and therefore, could not be a terrestrial phenomenon; as was supposed at the time.

Brahe was one of the most prominent scientists of his time, and it was with the help of his observations and stellar catalogs that modern astronomy was grounded. The fame of the Danish astronomer is not only due to his work in the field of science: it is partly linked to his personal life. In the year 1566, Brahe lost his nose in a duel while at the University of Rostock, and he had to wear a metal prosthesis for the rest of his life. The death of the astronomer in the year 1601, is still surrounded today with mysteries and legends. According to one of the oldest stories, he would have burst his bladder at a banquet because he refused to get up from the table during the meal (and at the time it was considered a gaffe). In the year 1901 there was an exhumation that revealed traces of mercury in his hair, and this raised the hypothesis of accidental poisoning, or not. The suspicions of poisoning fell on his assistant Johannes Kepler, who became a great astronomer after making use of the data collected by Brahe. Another possible culprit was King Christian IV, who would have ordered the death of the Dane for having an affair with his mother. Today, Tycho Brahe's body is buried in the Tyn Church in Prague.

- Johannes Kepler -

Johannes Kepler (Württenberg 27.12.1571 - Regensburg 15.11.1630) was born into a poor family and in the year 1584 he was admitted for free in the seminaries of Adelberg, but also studied in Weil, Leonberg, passed the seminaries of Adelberg and Maulbroun, where he obtained his master's degree in Tübingen in 1951. Strangely enough, while studying in Tübingen, he was initiated into astronomy by a teacher named Michael Maestlin, but this was a fervent Copernican, this would be his friend for all life. Johannes Kepler obtained the respect of the greatest astronomers of the time in his first book work in the field of astronomy, his books and called Prodromus dissertationum mathematicarum continens mysterium cosmographicum (First mathematical dissertations on the mystery of the cosmos), in 1596. Kepler acted like professor of mathematics in Graz since 1594, and in the year 1600, the year in which he was removed because he was Protestant. Brahe was already familiar with Kepler's first work, and when he left Graz, he went to Prague to find the Danish astronomer. When he arrived at the court of Rudolph II, he was elected as Brahe's assistant and mathematician. It did not take long and Brahe passed away, so Kepler succeeded the previous astronomer and inherited his notebooks. Kepler relied heavily on Brahe's observations, he was very good at it, indeed Tycho Brahe was the first astronomer to check his observatory instruments every day. Although Brahe made excellent observations, he believed in the heliocentric theory, which was the first to be taught to Kepler; however, he disagreed.

Kepler began to study Brahe's notebooks thoroughly, and it was after studying the orbit of Mars that he wrote another successful work: New Astronomy; of 1609. In this work the first two laws of planetary motion appear for the first time and first announced that the orbits of the planets in ellipses were not circular as was believed until then. He stated: "The orbits of the planets are ellipses in which the Sun occupies one of the foci" and also "the areas covered by the vector ray that joins the sun to the planet are proportional to time." This work immortalized his name. Kepler also recorded his name in the field of physics. In the year 1611, he published a work called the Dioptric, which became the most important work in the field of physics published before Newton's Optics. King Rudolph II passed away and Kepler had to leave, traveled to Austria and King Mathias commissioned him as the mathematician of the Austrian States in 1612. He published several works on various celestial bodies: Mercury, Jupiter satellites, comets and other so many. In the year 1619 he enunciated in Harmonices mundi: "The squares of the times of revolutions are proportional to the cubes of the great axes of their orbits."

This would be the famous third law of planetary motion, and the three laws are still used today. Kepler then devoted himself to working on the Rodolphinae Tables and published them in 1627, which most accurately indicated the positions of the planets. In the year 1628, he moved to Silesia and began working with ephemeris and in the year of his death he published the Admonnitio ad astronomos (Advice to astronomers).

- Galileo Galilei -

This is one of the most difficult biographies to be summed up in a few lines, because the guy was simply a genius. Galileo Galilei was born on 15 February 1564 in Pisa, Italy. He was a man of many functions: he was a writer, astronomer, inventor, teacher and physicist. There is a range of scholars who call it the title of first physicist. He began writing early, his father was a famous Florentine musician, and from the very beginning he presented the operas, and Galileo wrote some works on Dante and Tasso. Galileo studied in Florence and Vollombrosa, from 1574 to 1581, was when he entered the University of Padua to study philosophy and medicine. His father was very proud that his son would one day become a doctor; something that would never happen. Instead of studying the traditional subjects, he left everything aside to apply to mathematics. His gift was recognized and when his father discovered that he did not strive to graduate and wanted to bring him home, his teacher named Ostillio Ricci was mediator between Vincenzo Galilei and Galileo Galilei. It was not easy, but he succeeded and Vincenzo agreed to pay for his son to study for another year, but after that time he would have to go it alone. Time passed and in the year 1585 he left Padua for once, without having graduated. But like I said, he was good at math and he also had friends, good friends. Then in the year 1589, his friends got him an appointment to teach mathematics at the University of Pisa. It is interesting to remember that four years before he had left without a diploma. Now he taught at the side of his old masters.

However, teaching did not occupy him all the time and he could deal with his mathematical problems. But there he gained little and was always contradicting the ideas of Aristotle: his contract of three years passed and was not renewed. In December 1592 he began to teach in Padua, there he would conquer friends whom he would take for the rest of his life.

1609. The story of a Dutchman who had invented an instrument capable of increasing what the human eye could not see came to the Italian's ears. Galileo managed to get an instrument from Zacharias Janssen and perfected it. He perfected it to see and see: the stains of the Sun, the craters of the Moon, the phases of Venus, the rings of Saturn (which he called arm, because he could not identify them), the true composition of the Milky Way. These discoveries came to light in 1610. Galileo published a book that would electrify the European intelligentsia; Siderius Nuncius (Messenger of the Stars), a work that would bring unprecedented revelations of the cosmos. The book described the lunar reliefs, and this statement dethroned one of Aristotle's main ideas, which were all immaculate outside the earth; including the Sun and the Moon. The book still contained information about Jupiter's four satellites, Saturn's "arms" and the true nature of the Milky Way. In this work Galileo began to take advantage of the theory of Copernicus. However, Copernicus's theory ran headlong with the ideas that the Catholic Church propagated. Galileo got many people who opposed his ideas, but also people who stayed by his side. One of those people who showed him support was Johannes Kepler, who was considered one of the greatest astronomers in all of Europe. Kepler sent him a letter in the year 1610 encouraging him in his discoveries. When Kepler obtained a borrowed telescope, he looked at Jupiter's moons for several months and published a booklet confirming Galileo's observations. Another book of the Italian thinker began to reverberate: History of Stains to

Sun Accidents, where again he tried to show the mistakes of Aristotle. The first book passed through the Inquisition without many barriers, but when this second was published, the Holy Inquisition thought it was too much. Galileo Galilei did not see himself in bad linen at once, but was summoned to Rome in 1611 and had to defend himself against the charge of heresy. He did not get convicted, however, led an informal rebuke of the Inquisition and in the year 1616, had to sign a decree declaring the heliocentric system to be a mere hypothesis. In the year 1623 he published a book called Saggiatore (Experimenter), where he fiercely fought Aristotelian physics and established mathematics as the foundation of the exact sciences.

In the year 1632 he published the book "Dialogo di Galileo Galilei sopra i due massimi sistemai del mondo tolemaico e copernicano." (Dialogue on the two highest systems in the world: Ptolemaic and Copercano.) The book had a simple writing but contained tenebrous information. Galileo used his observations, his experiences and his convictions as a point of support and changed the world of place. He used to occupy the whole of the universe, now he was no more than a single planet, putting the Sun in its place.

Galileo was a man ahead of his time, and not even the greatest forces of the day (scientists, philosophers and even the Holy Inquisition) were able to stop the impetus of this fabulous genius. This man, who traded medicine for mathematics, and made a tool to unravel the world of physics and astronomy, establishing new principles, demystifying legends and denying theories, alone caused a revival impulse that was the most important one ever had been in the history of science. But he paid dear, very expensive.

At the end of the year 1632, Galileo received a papal order that he should go to Rome, where he would be tried, since he had disobeyed the order of the Inquisition to never speak again of the movements of the bodies. The Italian thinker was in poor health and for months the trip was delayed. In the end, the Pope gave him an ultimatum: "either you come immediately or I have you chained." In January 1633 Galileo left Florence and because of his weak illness he had to stop on the road, arriving at his final destination only on 13 February. He did not go straight to the jail, stayed in the embassy of Tuscany. There the Italian remained for two months, perhaps the Inquisition wanted to afflict him. On Tuesday, April 12, 1633, Galileo was arrested and was interrogated the same day. The inquisitors were picking up and Galileo argued. The interrogation was over and the

Inquisition kept the thinker in custody for two more weeks, not in jail but in apartment in the Inquisition building itself.

They wanted Galileo to confess and he confessed. On Wednesday, June 22, 1633, the Dominican Convent of Santa Maria Sopra Minerva, in Rome there was Galileo, in white robe kneeling before seven cardinals-inquisitors and reads, a confession already ready:

"I Galileo Galilei, son of the late Florentine Vincenzo Galilei, seventy years old ... I swear that I have always believed and will continue to believe in everything that the Holy Catholic Church postulates.

I was judged highly suspect of heresy for having believed and held that the Sun is at the center of the Universe and does not move, that the Earth is not the center and moves. [...]

With sincerity and true faith, I abjure, curse and abhor the aforementioned errors and heresies ... I swear that from now on I will never say or affirm, orally or in writing, anything that might attract such suspicion about me. "

According to legend, Galileo had completed the text like this: "Eppur si muove. (However, it moves)" But that is only a legend. The Inquisition banned Galileo from publishing anything and sentenced him to house arrest in a village called Arcetri. In the year 1634 his fiat Maria Celeste died and Galilei, who already suffered from serious health problems went into depression. The end of the year 1634 arrived and Galileo was good enough to resume his life. Although the weak one was blind in one eye, he went back to work. He spent two completing the book "Discorsi i desmostrazioni matematiche in due to due nuove scienze" (Mathematical Theories and Evidence on the Two New Sciences), written all in Italian, except for the mathematical theorems, which were written in Latin. Book with simple writing and was the same uncle as the previous one: a dialogue between three people, the same three from the previous book. This would be Galileo's scientific testament, and he himself said: "Now the door is open for a new reflection, rich in endless and admirable conclusions, which in the future may exercise other creative minds." The book was published in 1638 in Leyden, The Netherlands and was considered as the fundamental work of the dynamics. Galileo cannot see the copy of his work, for he had been blind. In the year 1638 he employed a young man named Vincenzo Viviani, and he went to live and work as his secretary. In the autumn of 1641 Galileo was bedridden, with palpitations in his heart and severe pains in his kidneys. On January 8, 1642, Galileo died in Arcetri. The grand duke of Tuscany asked for permission for a funeral homage and a monument in marble; but both tributes to the Italian thinker were denied. The body of Galileo Galilei was buried with modest ceremony in a crypt near a secondary chapel of the Church of Santa Croce in Florence, it was only in the year 1737 that the Inquisition allowed the mortal remains of the Italian to be transferred to the main part of the Church. They were placed in a monument of marble and bronze, that was financed by Vincenzo Viviani, who lived until the year of 1703 and was buried next to its friend and boss.

Galileo represents a very great importance in the history of humanity. In addition to everything he discovered, its undermined Aristotle's prestige. He wrote very important treatises, always polemical and didactic, incisive, ironic, in simple style sometimes in
Italian, sometimes in Latin. On September 12, 1982, Pope John Paul II visited the University of Padua where Galileo studied and was a professor, and withdrew accusations of heresies made by the Inquisition almost 350 years earlier. But

it was only in 1992 that the church considered him as a brilliant physicist. The Holy Catholic Church had to wait 360 years to redeem itself in the face of such a grave error: to support a pagan thinker and to condemn an experimenter.

Isaac Newton

The year 1642 led to the most celebrated astronomer and thinker in the world, but this year also brought a gift, another intellectual, another wise man: Isaac Newton. At first glance it seems a coincidence. Although Newton focused on mathematics and physics, he contributed a lot to astronomy. The English mathematician was a great genius, there is a statue at Cambridge, which was erected in the year of his death with the following words: "He surpassed humans with the power of his thought." He really did, gave order to what was once chaos. If the Universe were a clock, he would be the watchmaker, he dared to explain the workings of the Universe. He relied on reason and his model became the model of all sciences and forms of knowledge.

Isaac Newton was born in Woolsthorpe in the Christmas of 1642. There is a legend that every child born on Christmas Day is doomed to success, it turns out that Newton was not born at Christmas, England only adopted the Gregorian calendar years later. Newton studied at Trinity College, and in 1665 he received a bachelor's degree and still made discoveries in what we now call Newton's binomial. A short time passed and he began to work on differential calculus. In 1668 he built the first telescope for reflection. His studies in the field of astronomy took place from the year 1666, where he concentrated his attention on the problem of the movement of the stars, where he tried to sketch the important conclusions that his predecessors had arrived. The problem of the movement of bodies thrilled the scientific elite of the time. In fact, he had already solved this problem many years before. In 1648, a friend of his would have visited him, Edmond Halley visited him at Cambridge, and there he learned of a treatise which the scientist had written about the movements and, according to Halley himself, this was one of the most complete he had read. But there was a big problem, Newton's personality. Newton was arrogant and antisocial. He liked to find out things, but he preferred to keep his findings to himself, he repudiated the fame and company of other people. According to reports, he only fell in love once, and perhaps has only been related to a single woman. She was averse to criticism and isolated when her ideas were being debated. His work on the movements was only exposed to the academy in 1685.He returned to Cambridge in 1686 and began to write his (Mathematical Principles of Physical Science 1686-87). The work known as Principia, was composed of three parts: the first was entitled De motu corporum (The body movement), was completed in April 1686 and was presented to the Royal Academy on the same day; the second was completed in June 1687, this being an extension of the first, and in it Newton dealt with the movement of bodies in the resistant means; called De sistemate Mundi (of the World system), the third part offers a mathematical vision and explanation to the problem of the organization of world systems, preceded by philosophical considerations about the rules of reasoning.

The number of researches and discoveries of the English mathematician is very large. He introduced the infinitesimal calculus, perfected the manufacture of mirrors and lenses, was the inventor and maker of the first reflecting telescope, discovered the laws governing the phenomenon of tides, the laws governing movement and the force that holds the bodies together - or away. In the year 1701, he was elected deputy and moved to London. In the year 1703, he was elected president of the Royal Society of Sciences and was re-elected until his death. It was at this time that a startling dispute began: when Newton published his work on differential calculus, a German named Leibniz claimed to have created it first: it was proved that Newton created them first, though the two had come to the same conclusions independently. Newton began working at the Mint as director, appointed by Queen

Anne. In the year 1705 he was appointed under the title of Sir, the first scientist to receive such an honor.

In the year 1725, pneumonia struck over English, followed by gout. Newton still remained for two years carrying out his activities. On March 20, 1727, at the age of 85, he died. His funeral was a grandiose event: his coffin was carried by six members of parliament, was taken to Westminster Abbey; where it rests until today. It is interesting to remember that another brilliant scientist was mistreated, condemned, and defamed by the Holy Church, only 85 years earlier. Newton, by contrast, was the first scientist to be buried in an Abbey, also the first scientist to be buried in the presence of a pope.

The laws created by Newton are part of mechanics, which today is the part of physics that studies the movement of bodies, forces, energy and work. Certainly, there were no other scientists who expanded human understanding as the English scientist. Newton was responsible for taking a great leap forward in human thinking. Interesting to remember this phrase that was also put on the moon, because before we stepped there, he was responsible for preparing the way. Before him, the Moon was only a part of the sky, it was not subject to our laws and not even to our world and after him, it was a simple Earth satellite, kept in orbit by mutual force. Only after that we mere mortals could begin to understand how the gears that move the Universe work. It is interesting to emphasize that all this was only possible because Newton contributed in history at the right time, was under the shoulders of giants. It was only after Copernicus and Galileo broke the prejudice of the human mind, breaking all the restrictions of the medieval era. His greatest feat was to create a law that applied throughout the Universe. This Newton's assumption was not entirely correct, but it was audacious to the point of changing human thought, for from then on, scientists came to believe that everything that happened in the Universe could indeed be explained correct, but it was audacious to the point of changing human thought, for from then on, scientists came to believe that everything that happened in the Universe could indeed be explained.

Another Englishman and also friend of Newton: Edmund Halley. He was interested in a particular branch of astronomy: comets; He calculated orbits of comets, studied several of them, and even predicted the appearance of a new comet for the year 1758: it appeared and became the most famous comet in history: Comet Halley. What's more, Halley (the scientist, not the comet) has shown that just as the sun, moon, planets, comets do not travel without destination, they follow huge elliptical trajectories around the Sun, sometimes moving too far from the luminous star.

It was after the evolution of the observation equipment that the astronomical records became more and more precious. Still in the eighteenth century, an attempt was made to estimate the distance from Earth to the Sun by measuring the parallax of Mercury and Venus as they crossed the Solar disc. It was on this occasion that they made the discovery that the planet Venus had an atmosphere: when passing in front of the Sun, one could see a bright border around the second planet. In the year 1781, the German astronomer William Herschel found a celestial body, which he thought was a new comet. Over months of observation and after reviewing his calculations, Herschel confirmed that he had discovered a new planet, known today as Uranus. In 1783, Herschel discovered two new satellites on his new planet: Titania and Oberon. He also took a gigantic step toward a better understanding of the Universe: he discovered the binary stars, and showed that the same law of gravity that applied to the planets of the Solar System also applied to these stars. The distribution and catalog of stars in the night sky led Herschel to conclude that the Milky Way was a disco-shaped island universe, and the Sun was in the central position. Today we know that most of the nebulous objects he observed were other galaxies, other objects were gaseous nebulae; it is also known that the sun does not occupy the center of the galaxy, but rather one of the arms. After the discoveries of William Herschel, the biggest challenge for astronomers was to determine the distance of

stars and nebulae. When the spectroscopy was invented and developed (Kirchhoff and
Bunsen), discovered that the stars were formed by the chemical elements that were present on Earth. In the year 1842, the Austrian Christian Doppler established that the sound waves perceived by an observer change of frequency when the speed of the source changes with respect to the observer. Hippolyte Fizeau extended this phenomenon to the light waves in 1848, and this became one of the pillars of the cosmological models of the twentieth century. It was William Huggins (1824-1910) who first used the Doppler principle: he measured the radial velocity of the star Sirius and found that it was moving away from us. In 1917, the American astronomer VM Slipher (1875 - 1969) demonstrated that several spiral galaxies were moving away from us at great speeds. It was another American astronomer named Edwin Hubble (1889 - 1953) who took a major step in the study of distances: he observed the nearby spiral galaxies and discovered the variable stars (called
Cepheids). They allowed us to use a new method to determine distances. Albert Einstein
(1879 - 1955) was one of the greatest theoretical physicists of all time. he created the concepts that were, practically, the most important foundations of all current astronomical models. He dethroned Newton: he said that gravity was not a simple interactional force, but it was a consequence of the deformation of space, provoked by the presence of matter. He went beyond: space was not absolute, for he could never disassociate himself from time; thus, creating the concept of space-time. It was at this time that Einstein made the most tragic mistake of his entire career: he wanted to prove that the Universe was uniform, static and unlimited. His theory showed that it was not quite so, but in order to do so, he introduced a cosmological constant, which would be a simple force of repulsion between galaxies. That same year, the Dutch mathematician William de Sitter came up with a solution to Einstein's equations but without using the Cosmological

Constant. In the models proposed by Einstein and Sitter, the curvature of space-time would be positive and the Universe open. Thus, Sitter diverged from Einstein proposing an expanding universe. In 1957, the Russian Soviet Union launched Sputnik, the first artificial satellite into Earth's orbit; A year later, using the Explorer I satellite, James Van Allen discovered the Earth's radiation rings: it is the discovery of the magnetosphere. In 1961, Yuri Gagarin made the first manned flight to space, two years later, Valentina Tereshkova was the first woman to go into space. In July 1964, Neil Armstrong and Buzz Aldrin stepped on Lunar Ground. Ten years later another important discovery in astronomy occurred: Linda Morahito used the Voyage I imagery and discovered erupting volcanoes on Io, a moon of Jupiter. It was in the year 1987, Ian Shelton discovered the first Supernova visible to the naked eye since 1604.Ten years later another important discovery in astronomy occurred: Linda Morahito used the Voyage I imagery and discovered erupting volcanoes on Io, a moon of Jupiter. It was in the year 1987, Ian Shelton discovered the first Supernova visible to the naked eye since 1604.Ten years later another important discovery in astronomy occurred: Linda Morahito used the Voyage I imagery and discovered erupting volcanoes on Io, a moon of Jupiter. It was in the year 1987, Ian Shelton discovered the first Supernova visible to the naked eye since 1604.

2017. The clash of two black holes shows scientists what Albert Einstein had predicted 100 years ago: gravitational waves.

2018. This year, science and the world lost one of its greatest collaborators:

Stephen W. Hawking.

Today we have beautiful photographs and many and many discoveries in the field of astronomy because of a 1990 feat: the US launches the Hubble Space Telescope, named after the astronomer Edwin Hubble. A year later, Alexander Wolszcan discovered planets in a pulsar orbit, these were the first exoplanets (planets outside the Solar System) discovered.

The first exoplanet orbiting a normal star was discovered in 1995 by Michel Mayor and Didier Queloz, who was baptized with 51 Pegas B.

And so on. To this day the objects of observation and measurement are perfected as each new discovery is achieved, it is our advance towards the future. With each new discovery, we take a step closer to understanding who we are, where we came from, and where we are going.

THE BEGINNING OF THE UNIVERSE

The part of astronomy that studies the origin, structure and evolution of the
Universe, or at least part of it is cosmology. Astronomy itself is a very broad science and for this reason has many subdivisions. The so-called fundamental astronomy is the part that studies the position and movement of the celestial bodies. Astrophysics deals with the construction, physical properties and evolution of the stars. Stellar astrophysics (a division subdivision) is associated with the composition, formation, birth, growth and death of stars.
Dust, gases and forms of radiation between stars are studied by interstellar astrophysics. Galactic astronomy studies our galaxy other than the extra galactic, which studies how galaxies come together to form larger systems and also study the other galaxies. Cosmology, as we have said above, studies the origin, structure and evolution of the Universe. Planetary astronomy analyzes the planets (not very easy task, define what a planet is), comets and asteroids. There is still instrumental astronomy, where engineers, astronomers and computer professionals develop devices that allow the study of celestial stars. For cosmology, the Universe is the space with matter and energy that constitute it.   Several famous names dedicated their lives in the studies of science and discoveries of the origin of the Universe. In chronological order we can cite:

1914 Albert Einstein enunciates the Theory of General Relativity, showing the equivalence between matter and energy; Creating the most famous equation in the world: $E = M.C^2$. 1917 - The Dutch astronomer Willem de Sitter theoretically demonstrates that the Universe is expanding. 1927 - Belgian astronomer Georges Lemaître suggests that initially all matter in the Universe was concentrated in one place: the cosmic egg or primordial atom. 1929 - Edwin Powell Hubble, based on his observations, enunciated his famous law according to which the speed with which a galaxy moves away from us is related to its distance to us, and therefore with time. This was the first evidence of universal expansion.

1950 - Hermam, Gamov and Alpher propose the Big Bang theory as an explanation for the event that gave rise to the Universe.

1965 - US physicists Arno Penzias and Robert Wilson detect the background cosmic radiation, equivalent to the radiation emitted by a black body of a temperature of 2.7 Kelvin, this finding corroborates with the Big Bang theory.

Hubble's 2nd Law

Astronomer Edwin Powell Hubble (1889-1953) in 1929 gathered enough elements to conclude that the speed with which a galaxy moves away from ours is directly proportional to its distance from ours. Mathematically: $V = Hd$; where V is the spacing velocity of the galaxy in question; d is the distance between the galaxy considered and ours and H is the Hubble constant. Hubble's law suggests to us that all matter that exists and is expanding at a given instant may have been all concentrated in one place. This place we call the cosmic egg, or singularity.

Big Bang Theory

It is possible the existence of many Universes. Ours arose about 13.7 billion years ago in a great explosion that time, energy and all matter was created. This name was given in a contemptuous way by some antagonist of this conception, in repudiation to the theories that pointed in this direction.

According to this theory, about 13.7 billion years ago, an enormous amount of energy was gathered in a single point of the Universe in a sphere, called cosmic egg or singularity. At a given instant (Time = 0), all of this rapidly expanding energy created the Universe, which expanded and cooled uniformly.

In detail:

The Universe, after the Great Explosion, began to cool and expand, and as it expanded, this cooling produced so much energy that it began to stabilize. Protons and neutrons were "created" and stabilized when the Universe had a temperature of about 100,000 million degrees, about a hundredth of a second after the explosion. The electrons had a great energy and interacted with the neutrons, which initially had the same proportion as the protons, but due to these shocks the neutrons became more protons than vice versa. The proportion continued to drop as the Universe continued to cool. So, when the Universe was 30,000 million degrees (one tenth of a second later) it had thirty-eight neutrons for every sixty-two protons and twenty-four for sixty-six when it was ten.000 million degrees (one second after the Big Bang). The first to appear was the nucleus of the deuterium, almost 14 seconds later, when the temperature of 3,000 million degrees allowed the neutrons and protons to remain together. Until these nuclei could be stable, the Universe took about three minutes. It was when this glowing ball cooled, reaching a temperature of 1,000 million degrees.

For our theory here to be valid, scientists had to understand where all this energy was created, since what we see is very little. That was when the postulate for dark energy was created. It is named because it is not seen and does not interact with any of the nuclear forces.

What determined the successive transformations of energy was the rapid reduction of temperature. The energy materialized in the form of particle (quarks) and antiparticles (antiquarks). Matter and antimatter annihilate, generating an enormous amount of energy in the form of photons, obeying Einstein's equation: $E = M \cdot C^2$, that is, matter is imprisoned energy, and energy, liberated matter. The result masses. Speed of light squared shows that there is an enormous amount of energy in each piece of matter. The excess of matter in relation to anti-matter gave rise to the Universe in which we live today.

It was George Gamov (1904-1968), a Russian physicist who naturalized himself American, together with Ralph Alpher and Robert Herman began in 1940 the first conjectures for the elaboration of a theory of the expanding universe. He relied on Edwin Hubble's experiments in 1929 that galaxies are moving away. This would be evidenced by the Redshift (redshift) displayed by cosmic clusters when viewed from Earth. This reddish aspect would be a manifestation of the Doppler effect applied to the light, verifying when a luminous object affects with great speed of a fixed observer.

Other evidence corroborating the possibility of the Big Bang occurring is the gradual cooling of the Universe, which today has a temperature of 2.7 K, much smaller than the corresponding value in the past ages.

In 1965, two American physicists at Bell Telephone Laboratories in New Jersey,
Arno Penzias and Robert Wilson were testing an ultra-sensitive microwave detector
(microwaves are like light waves, but with a frequency of only 10 billion Hertz). Penzias and Wilson were worried when they saw that the detector, they were testing made more noise than it should have. The noise did not seem to come from any particular direction. After some tests, they saw that the noise was coming out of the atmosphere. It was the same day or night, during the year. That sizzle they heard was the

electromagnetic signals in the microwave range, permeating space in all directions and were probably produced 380,000 years after the primordial instant: The Big Bang.

THE BIG BANG CHRONOLOGY

a)   Moment T and T = 0: initial moment when the Big Bang occurred. It is possible that there was something before him, but as far as we are concerned, the events that preceded him cannot have had any consequence and therefore should not be part of a model of the Universe, so we must isolate them and consider them the time begins with the Great Blast. The density and curvature at this point is infinitely high. There are still no physical or mathematical tools that can study this moment, which is treated as a singularity in the evolution of the Universe.

b)   Time interval between T = 0 and T = 10^{-43} second: what occurred in this time interval is pure theoretical speculation, with no possibility of physical proof. It is believed that in this space of time space and time have arisen, at an estimated temperature of 1032K.

c)   Time interval between T = 10^{-43} and T = 10^{-35} seconds: it was in this short time interval that the quarks and antiquarks annihilated giving rise to radiation in the form of photons. The number of quarks is higher than that of antiquarks, so that matter is left in the form of quarks, where the universe we live in today begins. The Universe begins to cool and passes to a temperature of 1027K at T = 10^{-36}.

d)   Moment T = 10^{-36} s: the remaining quarks of the annihilation process began to merge giving rise to protons and neutrons.

e)   Instant T = 10⁻⁶ s: Quarks merge is complete, quarks disappear. Protons and neutrons can transmute to each other and do not coexist with electrons and photons.

f)   After the time T = 15 at T = 500s: the fusion reactions of the nuclei occurred. 10 percent of the hydrogen nucleus is transformed into helium; One thousandth percent is transformed into deuterium (heavy hydrogen) and less than one millionth of a percent is converted to lithium. At this stage, the Universe finds itself as a kind of luminous and non-transparent soup (like a gas lamp). The light did not allow the agglutination of matter.

g)   At time T = 400 thousand years: the temperature dropped to 3,000 K and the ionized plasma was neutral. The sky became transparent and dark, as it is today. At this same time there was a recombination: neutral atoms formed, releasing photons that gave rise to cosmic background radiation in microwaves.

h)   100 million years ago: by this time the temperature had already dropped to a few thousand degrees, and not enough electrons and nuclei to cope with the electromagnetic attraction between them, would have melted to form atom. The Universe continues to expand and cool, but in regions that are slightly denser than the average, expansion would have slowed due to the extraordinary gravitational pull. This would eventually stop the expansion and in some regions would trigger the start of a new explosion. When they collapsed, the gravitational pull of this matter out of these regions could start to spin slowly. As the collapsed regions became smaller, they would begin to acquire a speed increase in their rotation. Later, when the region became small enough, would be spinning fast enough to balance the gravitational pull, forming different rotating protogalaxies. Other regions that did not go into rotation would have become spiral objects, called elliptical galaxies. In these, the collapse would be interrupted because certain parts of the galaxy would be spinning with stability around its center, although the galaxy itself did not spin. As time passed, the hydrogen and helium gases in the protogalaxies would divide into smaller clouds, which would collapse under the effect of their own gravity. As they contracted and the atoms within that cloud collided with each other, the temperature of the gas would increase until eventually it became hot enough to trigger nuclear fusion reactions. This would convert hydrogen into more helium, and the value expelled would increase the pressure, thus interrupting the continuation of the contraction of the cloud. For a long time, this transformation of hydrogen into helium would continue: the stage at which our Sun finds itself, making its nuclear fusion and radiating energy in the form of light and heat. Thus, formed the proto-stars. Larger masses need more heat to counterbalance their gravitational pull stronger, causing nuclear reactions to consume hydrogen in a shorter time (about 100 million years). Thus, they would contract slightly, and from then on, as they warmed, would begin to transform helium into heavier elements, yet not enough energy would be released and would explode. The future of this star will depend directly on its mass. When it explodes, it can transform into a neutron star or a black hole. It could also explode and become a supernova, which would illuminate its entire galaxy, then gradually fade away, until only one super dense core remains.

In the case of stars like our Sun, when its mass is extinguished, which will happen in five billion years, its explosion will result in a planetary nebula. In its final stage it will return to the Universe all the remaining elements, which will later be used in a new generation of stars.

And the unanswered questions?

As we know, the Big Bang theory is not the only one that tries to explain the origin of the Universe, I will explain some others later. After all, the Big Bang Theory is the only one that explains homogeneously the beginning of the Universe as it is known. For many times it has been revised and many other times have tried to overthrow it. Even so, it is the most accepted by scientists; even though some questions remain unanswered.

Some of them are:

a) Why would the initial Universe have been so hot?

b) Why is the temperature of the cosmic background radiation the same even if it is observed in regions and at different times?

c) Why would the Universe have begun to expand at a rate so close to the critical point, reporting models that have collapsed again, from those that have been expanding forever, so that even now it continues to expand to the same critical reason?

There have been other questions, such as why the Universe is so uniform on a large scale or why it seems so equal in all points of space and in all directions. The new findings show ways that can answer these questions. You may also as long as you read this book, these questions have already been answered, or other questions such as the one I have presented and we have not been able to answer so far appear. As our equipment advances, it is likely that the cosmology in a few years' time will be able to respond; and if it is not, we should create a new theory.

The New Explanations of Cosmic Genesis

After many and many years of study, our tools for the study of the beginning of the Universe have developed on an arithmetical scale: observation equipment here on Earth and in space, quantum field theory, high energy elementary particle physics, the refinement of inflation theory, new discoveries ... Anyway, many tools have been added to help us in this rather difficult task: to find out what happened at the beginning of everything. With this, many researchers began to take an interest in the primordial Universe. It was initially thought that a detailed understanding of the very high density and energy physics of the primordial universe was essential to understanding what happened. It did not take long to verify that it was quite the opposite, in fact, it was much more than that.

Scientists dared in the imagination: to understand how the properties associated with space and time had arisen, how the Universe was created. Some are bold, or at least innovative. As theories are less accepted, I have removed the physical and mathematical part for a better understanding, however, I make it clear that a theory is not only a good idea in the head, but with a lot of physics and mathematics, which is totally necessary for prove it.

The Universe that reproduces itself

The self-reproducing universe theory says that the Big Bang began as a microscopic quantum fluctuation that occurred somewhere earlier than ours. In the same way, our Universe may be "pregnant" with other Universes. That is to say that at any moment other events similar to the Big Bang can occur, only this time in our own Universe. The problem is that these explosive births would be viewed with great difficulty or even, they might not even be seen at all.

Chaotic inflation

This theory was proposed by Andrei Linde, a researcher at the Lebedev Institute of Physics in Moscow, Russia. For Linde, the cosmos is a self-reproducing entity, which exists eternally, and which is divided into several mini-Universes, some of which are much larger than the observed portion of our own Universe.

According to Linde, the laws of low-energy physics and even the dimensionality of space-time may be different in each of these Universes. In Linde's theory, the quantum field that gives rise to the Universe is not smooth on a microscopic scale, but instead resembles a chaotic, inhomogeneous "space-time foam." In some regions of this foam, the energy density could be as high as 10^{-33} grams per cm³, that is, 125 orders of magnitude higher than that of the Universe visible today. Andrei Linde further claims that other Big Bangs could get anywhere in this foam and remain totally disconnected from each other forever. It could even occur that an infinite number of universes originated in these fields, without one interfering in the other.

Similarly, our own ordinary space-time could be chaotically "sparkling" on the very small scale of 10^{-33} cm. It could also give birth to other universes. The creations could thus form a cascade, one from the other, each producing perhaps many other universes. Some of these universes suffer collapse at birth, while others are expanding at high inflation. Andrei Linde states that in this cascade of multi-universes, our Universe is probably the first of all.

Eternal inflation

When American physicist Alan Guth expounded his idea of inflation theory, a flurry of variations on basically the same theme and at the same time were elaborated by several scientists, and so several names were given to a same theory, names such as "hyperextended theory", "eternal inflation", hybrid inflation, etc. Let us look at some of these ideas, beginning with "eternal inflation", which was proposed by Guth and Linde. According to the scientists, after about 10^{-30} seconds of inflation, half of the original region that was in the false vacuum state would have fallen into a normal vacuum state and in that process, half of the universe would have remained in a false vacuum state, which means he continued to suffer inflation.

Theoretical calculations show that the rate of this inflation, in fact, would have been much higher than the rate at which the false vacuum study fell. A simple reasoning says that if some areas were suffering inflation much more quickly than others were suffering decay, inflation would have overcome the decay. In addition, until now some part of the Universe should still be suffering the inflationary process. In conclusion: one that inflation is initiated, it is eternal.

Multi-universes

If Linde's proposal above (the theory of "eternal inflation") is correct, there were still some areas where inflation never really happened. In this case, we would have resulted in a multi-branched entity, fractal and gigantic: the multi-universe.

According to this theory, our Universe would have been born of a small bubble in space-time that suffered inflation from a pre-existing region. This region, in turn, has inflated from a previous region and so on. We could follow this line of reasoning backwards in such a way that the original birth of space-time could be useless to ask how it occurred. In this theory, our Universe would be just one component of the multi-universe, which continues to grow through a series of Big Bangs much longer than our small region in the entire multi-universe. And this process will be eternal.

This theory is very complex. In fact, it proposes that the laws of physics would probably be different in every branch of this immense tree. That means it could never be possible to understand how other parts of the multi-universe have worked so far. This would make it impossible to deduce a single unified general theory for all laws of physics, since they could, and certainly would be, different in every part created within the multi-universe.

No-Boundary hypothesis - The hypothesis without outline

In the year 1983, James Hartle and Stephen Hawking began working on cosmological singularity using a mathematical approach known as "path integral." They seek to calculate the problems that arose if the Universe had begun in many different ways. Thus, was born the proposal that became known as No Boundary Hypothesis, which is freely introduced as hypothesis without bound.

To understand this proposal, let's use a simple geometric figure: a cone. Let's imagine that this cone represents the evolution of the Universe. The time grows next to the cone, from its vertex upwards, while the space moves around the cone. Note that you move upward in the cone, from its apex to its base, so in the increasing direction of time, the width of the cone (representing the space) increases. The origin of time and space occurs at the point located at the bottom of the cone at the vertex. In traditional Big Bang theory, this is the uniqueness.

However, in quantum physics there is no such thing as a precise point, there is always an uncertainty associated with it. To visualize this, imagine that the point is rounded as if it were the door of the ballpoint pen, decreased to an infinitesimal size. This is exactly what the Hartle and Hawking trajectory integral predicts as the most likely setting for the birth of the Universe. Instead of the dimension of time (which grows along the side of the cone) begins at a discrete point, it emerges from the dimension of space (around the cone). And just as there is no point on the surface of a sphere where we can say that the sphere "begins," there is no distinct point on the rounded hemispherical bottom of the cone of space and / or time begins. The conclusion is not difficult: there is no starting point and there is no distinction between space and time, just as there can be no distinction between past and future.

Briefly: Hawking and Hartle proposed that cosmic inflation occurred from the rounded apex of a cone. There was, however, a problem with the theory: the calculations that the two scientists completed produced only closed universes. As the current observations suggest in plan, the model of creation of scientists leads to a Universe of a different kind from which we live.

There are still other names, Hawking worked with Neil Turok to create a Universe that would be a bubble inside another bubble. They "concerted" the theory "No Boundary Hypothesis", and "created" the Universe from a "cosmic pea", which they called Instanton. Paul Steiherdt, Romana Sundrun, Lisa Rundall, and many scientists worked to create a theory to explain 100% the creation of the Universe. For now, the one that answers the most questions are the Big Bang (the traditional!).

Dark energy

Five billion years ago, mysteriously, the Universe had an acceleration in its expansionary rate. The question the cosmologists tried to answer was: what could be fueling this expansion? The answer may lie in so-called dark energy, a repulsive force that counterbalanced the attractive axis of gravity, which was holding back the growth of the Universe. It was then that for the first-time scientists were able to measure the slowdown of up to 11 billion years ago. The chart shows a cosmic roller coaster that began its journey shortly after the Big Bang.

New measurements reveal a time when the Universe was climbing the chair, being restrained by gravity. To measure this deceleration, astronomers used the aspects of 50,000 quasars in the distant Universe made by the SDSS - Sloan Digital Sky Survey. Thus, they were able to assemble a three-dimensional map of the distribution and structure of the gas and matter clouds of the primordial Universe in the line of sight between the quasars and Earth, showing the evolution of their size.

The different contractions of matter and gas were generated by the so-called BAO (baryonic acoustic oscillations), marks left in the structure of the cosmos by the interaction between "common" matter, known as baryonic, as well as mysterious dark matter and radiation united in the Primordial universe after the Big Bang. Just before they split up in a process called triggering. Dark matter began to form "pockets" which, in turn, attract dark matter into the formation of stars and galaxies.

In addition to determining the overall expansion of the Universe, dark energy has long-term consequences on smaller scales. When one zooms in on the internal observable Universe, the first perception one has is the distribution in the web pattern of matter on cosmic scales: a series of filaments, several thousand million light years in length, intertwined with voids of size similar. These filaments are traits involved in the competition of cosmic expansion (and anything that affects it) and its own gravity. They did not detach themselves from the general cosmic expansion and established their own balance of forces. In our Universe, none of these actors in this tug-of-war is extremely dominant: if the dark energy were stronger, the expansion would have conquered and the matter would have spread rather than settle on filaments. If the dark energy were weaker, the matter would be even stronger than it actually is. Dark energy is much better known as the supposed agent of cosmic acceleration, a type of indefinite substance that exerts a kind of antigravity force acting on the Universe as a whole. There is also the possibility that this type of energy may have reduced the growth of galactic clusters about six billion years ago. What little is known is that, it helped to imprint the characteristic pattern of filaments (which is only observed at large scales). There is also the possibility that this type of energy may have reduced the growth of galactic clusters about six billion years ago.

The Evidence of Dark Energy

Supernova Explosions

In an expanding Universe, the galaxies move away at a speed that depends on the distance between them. Supernovae offer the possibility of measuring this effect: the redshift in its spectrum reveals the speed of the galaxies in which they inhabit and their brightness determines the distance of the expansion. It was discovered that billions of years ago, galaxies moved more slowly than was expected through a simple extrapolation of the current velocity of expansion. The speed of expansion of the Universe, therefore, would have increased from here to here - the hallmark of dark energy.

Cosmic Microwave Background Radiation

In the images of the cosmic background radiation contains stained regions whose apparent size reflects the geometry of space as a whole and therefore the density of the Universe. The value of this density exceeds estimates of the amount of matter (both common and exotic) in the Universe. Thus, a component not yet detected - the dark energy is a candidate - should make the difference. In addition, background radiation has been slightly altered by the gravitational fields of cosmic structures. The degree of this change is
a function of how much the velocity of the expansion has changed over time and corresponds to the effect predicted for the dark energy.

Galactic configuration

The galaxies are not randomly scattered in space. They are arranged in patterns, one of which resembles the stained regions of background radiation. This can be used to measure the total mass of the Universe and note the need for dark energy.

Gravitational lenses

The amount of mass can bend the light like a lens. If a light source is directly behind that cluster, it will form several images. The larger the size of the Universe - which depends on the amount of dark energy - more likely, this type of alignment becomes a weak lens is still able to deflect light at a small angle, which depends directly on its mass. Matter clusters grew over time and find the dark energy mark.

Clusters of galaxies

Observations of the frequency of X-rays show the evolution of the mass of the galactic agglomerates. Only with dark energy can a theory be developed to explain them.

Dark matter

For some years now, we have known that matter emitting electromagnetic radiation (therefore visible) is only an insignificant part of all matter that has a gravitational function. We know for certain that it exists only because we detect its effect under the luminous matter.

It all began in the 1930s, when astronomers found the first clues that at least 90 percent of the Universe mass does not emit, reflect, or even absorb light. The invisible material - which has been called dark matter - prevents galaxies from undoing on all sides, keeping the clusters of galaxies together. It also seems that dark matter may have played a crucial role in the formation and development as we know it today. If it corresponds to 90% of our Universe, and the final destination of our Universe depends on its mass; it can determine what the end will be like.

In the year 1933, Fritz Zwicky was examining Comma's Particle in the constellation of Comma Berenice, found that some discovered that some galaxies were moving at a very high speed. In fact, this is what happens: the galaxies of the cluster in Comma Berenice move so fast that all visible stars and gas in the cluster cannot keep the galaxies gravitationally attached to each other, according to the physical laws as we know them. However, the agglomerate remains intact

Zwicky, for the first time, concluded that some kind of unseen matter should exist there to provide the force of gravity lost. For many years, this idea was in the dark. Some scientists have tried to study galactic movements in more detail, theorizing that the reason for the existence of invisible material could disappear. But this did not happen, quite the opposite.

The last piece of the puzzle

Cosmologists had an elementary question: how did the Universe evolve from a uniform soup of elementary particles after the Big Bang to soon reach its current structure of clusters of galaxies and superclusters? Scientists cannot believe that visible matter came alone in the huge structures we see today, although our Universe is about 13.7 billion years old

To complete this cosmic puzzle, scientists have convinced themselves that the Universe contains a special type of dark matter, which was called cold dark matter, which moves more slowly and joins in groups faster than visible matter.

The theory of cold dark matter may be correct, but like all theories it has some flaws, but there are small disambiguates. This theory says that large galaxies should be surrounded by satellite galaxies, which in fact does not occur (at least it is not observed).

This theory either needs improvement or we need a better one. It may also be that these satellite galaxies are orbiting around us, but we have not yet observed them. Were they galaxies formed only by matter and dark energy?

Challenging Sir Isaac Newton

When Newton proposed the law of Universal Gravity, when we speak of the others, it was to be expected that - as indeed it happened - the planets (such as Mercury and Venus) orbited the Sun at a faster rate than the outermost planets like Uranus and Neptune. When we speak of the stars and make an analogy with our Solar System, it is expected that stars closer to the central bulge of the galaxy (if we make an analogy of a fried egg, galactic center, bulge would be the gem) orbited in a speed far higher than the peripheral stars (which are situated at the edge of the galaxy). But nature reserves surprises for us and what happens is different: peripheral stars have speeds as high as the innermost stars, and there is a question of interest: What is it that holds this star to the galaxy? With orbital velocity so fast, and with so little visible material as the second at the edges, why do not they escape the galaxy they inhabit?

So, to answer those questions, scientists have come to the conclusion that everything we see in a galaxy is only a small portion of the total mass of a spiral galaxy.

Completing the Universe

When the Universe is observed in any observed region, it appears to have the same appearance (when observed on a large scale), and this indicates that there should be a critical density in it, but when we observe the visible matter it is not sufficient to maintain this density (which, I have already said about it). But if we could count on more matter (and even if we could not see it), we would reach critical density: dark matter is a good candidate.

Stars

When we look at the sky at night, and the sky is clear, we observe a lot of speckles that glitter and that decorate our night sky. From the very beginnings of human intelligence, we look at the sky and try to understand what they are and why they exist. Those dots are magnificent thermonuclear plants scattered throughout the space-time of the Universe. The stars are hot bodies resplendent with gas, which are born in large proto-stellar nebulae
(giant clouds of gas and dust). There is an estimate that there are about
100,000,000,000,000,000,000. When we look at the sky at night in unarmed view, we can observe about 8,500 stars and when we use a telescope or telescope, we see other infinity of them. There are a variety of types, sizes, masses and temperatures: the diameter goes from 450 times smaller to about 1,000 times larger than the sun. The mass can vary from 0.25 to 50 solar masses. The color of a star is determined by its temperature: the hottest stars are blue and the coolest stars are red. As our sun has a surface temperature of 5,000 °C, it is in the middle of these extremes and has a yellowish color. The energy that is released by a star is produced by nuclear fusion in its central core. The brightness of a star is measured in magnitude - the brighter the star, the smaller its magnitude, which are of two types: apparent magnitude and absolute magnitude. The apparent magnitude is that which the object exhibits when viewed from Earth and absolute magnitude which is the brightness as seen from a standard distance of 10

parsecs (32.6 light years). The absorption rays are a series formed by dark lines which is the decomposed light was emitted by the star. The streak patterns indicate the presence of particular chemical elements, which allow astronomers to infer how the stellar atmosphere is composed. A chart called Hertzsprung-Russel measures the color and magnitude of a star.

The brightest stars seen from Earth

Common name	Apparent magnitude	Name of the constellation	Spectral class
Virus	-1,5	α Canis Majoris	A
Canopus	-0.7	α Carinae	A
Rigil Kentaurus	-0.3	α Centauri	G
Arcturus	-0.4	α Bootes	K
Vega	0.03	α Lirae	A
Capella	0.1	α Aurigae	G
Rigel	0.1	β Orionis	B
Prototypi ng	0.4	α Canis Minoris	K
Archenon	0.5	α Eridani	B
Betelgeus e	0.5	α Orionis	M
Hadar	0.6	β Centauri	B
Acrux	0.8	α Crucis	B
Altair	0.8	α Aquilae	A
Aldebarã	0.9	α Tauri	K

Antares	1.0	α Scorpii	M
Spica	1.0	α Virginis	B
Powellux	1.1	β Geminorum	K
Formalhault	1.2	α Piscis Austrini	A
Must	1.3	α Cygni	A

The Hertzprung-Russel Diagram

This is the Hertzsprung-Russel diagram. It is in him that we frame the stars according to their spectral type that roughly would be like we do with fingerprints to identify people: when we make the light of the star pass through a prism or spectrograph, the spectrum (or sequence of radiations) emitted by the elements, which helps to identify it, also helps to identify its absolute magnitude. The first to present the diagram were physicists Ejnar Hertzsprung (1873 - 1967) and Henry Norris Russel (1877 - 1957) in the first half of the twentieth century. This graph represents the stars by points: in the horizontal one is denominated the sequence of the spectral type and in the vertical by the absolute magnitude. The distribution is not uniform, most are concentrated along a line, which is called the Main Sequence, which begins with warm types O and B (known as the blue giants) and goes to the red dwarfs of the spectral types G, K, and M.

Spectral types

The analysis of the spectrum of numerous stars shows that there are many common properties in them, which may be almost all, distributed in seven main categories, which are called spectral type. Are they:

Type O: blue stars, very hot, its surface temperature can reach 30,000 K and, in its atmosphere, can be found ionized helium and heavily ionized metalloids;

Type B: bluish white stars, with a surface temperature of 12,000 K, and in its atmosphere found H and He neutrons and medium-ionized metalloids;

Type A: white stars. Surface temperature of 8,000 K. There was predominance of hydrogen on the surface, but there are calcium and ionized metals;

Type F: yellowish-white stars and surface temperature of 6,000 K. The hydrogen atmosphere has a strong presence of ionized metals, principal iron and uranium.

Type G: yellow stars, one of the Solar type, with surface temperature of 5,000 K and in whose atmosphere not found hydrogen, ionized calcium and molecular associations like CH and molecular carbon.

Type K: orange stars as the spectrum similar to those studied in the atmosphere, in addition to the constituent elements of the G-type study and still found the titanium oxide and much more than in all other types, neutral metals.

Type M: cold red stars, with a surface temperature of approximately 2,000 K, with large amount of titanium oxide in its atmosphere. But it could not be that easy and all the stars fit into these seven kinds. A few stars fit into this division and then it was necessary to create other types.

Type W: also called Wolf-Rayet stars, are warmer than type O, exceeding the 3,000 K marks and emitting rays are shown in their spectra. Their atmosphere may also be similar to those of type O stars, but they are usually richer in nitrogen or carbon and oxygen.

Type S: can also be called zirconium oxide stars, quite cold and their temperatures are no more than 2,000 K. Its atmosphere, as the name suggests, contains large amounts of zirconium oxide, but there may also be, in some cases, titanium oxide.

Type C: They are carbon stars. Cold stars that can be compared to stars of type K and M, in color and temperature.

In their atmosphere there are carbon compounds such as CN and molecular carbon.

To increase the difficulty a bit, each of these types can be further divided into 10 subtypes, so that a continuous sequence of spectral classification can be obtained. In this case, type F will contain the subtypes F0, F1, F2, F3 ... F8, F9 and the next type will be G0. The difference between subtypes is small, being type F9; more similar to G0 than F5.   Spectral analysis not only determines the chemical composition, but also provides other data of equal importance. The width of lines allows to calculate the speed of rotation of the star around its own axis: the unfolding of the lines in two or three components means the existence of a magnetic field, the intensity of which can be determined by measuring its distance between the components of the unfolded line. The Doppler effect, the physical law enunciated by Christian Johan Doppler in Prague: studying the stars in motion, the lines of their spectra do not appear exactly in the same positions as in the laboratory, but rather shifted to blue or red (blueshift or Redshift) as the star approaches or distances itself from our planet, the displacement being directly proportional to velocity.

Spectral classes of stars

Class	Colour	Surface temperature	Example
O	Red - White	30,000 K or more	Lambda Orionis

B	Blue - White	12,000 K to 30,000 K	Rigel	S
A	White	8,000 K to 12,000 K	Sirius	m all
F	Yellow - White	6,000K to 8,000K	Prototyping	m as
G	Yellow - White	5,000 K to 6,000 K	Sun	s
K	Orange	3,000 K to 5,000 K	Arcturus	sta rs
M	Red	Less than 3,000 K	Antares	

Stars with mass up to about one and a half times the mass of the Sun are considered as small stars. A star smaller than 0.5 Solar mass will never be able to start the melting of helium, even after the nucleus ceases to melt hydrogen. It simply has no mass required to exert sufficient pressure under the core. These are red dwarfs, such as the Near Centauri, some of which will live thousands of times longer than the Sun. Star formation begins when a region in a nebula condenses and contracts under the very force of gravity, where a huge gas globule and dust is formed. The proto-star begins to glow as the regions of matter in condensation within the globule begin to warm and glow. For the star to begin nuclear fusion that will turn hydrogen into helium, it needs to reach a temperature of about 15 million degrees Celsius, and this only happens if it has enough matter. In this process, the star begins to release energy and thus prevents it from contracting. Once she passes this troubled birth and begins releasing energy in the form of light and heat, it fits into the main sequence. A star such as our Sun, for example, can remain in the main sequence for about 10 billion years, until all the hydrogen in the core has been converted to helium. The resulting lump of the melt, now composed entirely of helium, contracts again and the nuclear reactions continue in a layer around the core; however, this process does not happen at all stars. If the star is totally connective (which it believes to be the case of the smaller stars) it will not have these layers circulating and will simply contract, until the degenerating pressure of the electrons interrupts its collapse and the star will directly become a white dwarf. If the star gets its nucleus stagnant, then it will have its nucleus surrounded by layers of hydrogen, which the star will attract. In this case, the core will become warmer enough that the helium, by nuclear fusion, transformed into carbon, while the outer layers expand, cool and shine with less intensity. The expanding star is called the red giant. When the helium runs out, the star's outer layers expand into a shell of gas called the planetary nebula. In other cases, some stars begin to melt the helium in superheated points of the nucleus, causing an unstable and irregular reaction, as well as a strong stellar wind thus, the star evaporates and leaves behind only a dwarf brown, without forming a planetary nebula.

Stars with intermediate mass

The stars that have mass between 0.5 and 10 solar masses, become red giants of two types:
(I) Branch stars of the red giants, whose layers are still fusing hydrogen into helium, while the helium nucleus is inactive. They reach hydrostatic equilibrium after the degeneracy pressure of the electrons is sufficient to counterbalance the gravitational pressure. (II) Stars of the giant asymptotic branch, which has a nucleus that passes through the fusion of carbon.
In any case, the accelerated fusion of the hydrogen-containing layer immediately above the nucleus causes the star to expand. This moves the upper layers away from the nucleus, reducing the gravitational force on them, and

expanding faster than increasing energy production, causing the star to cool, causing it to become redder than it was when it was in the main sequence.

A red giant is a major nonsequential major star, according to the Hertzsprung-Russel diagram. The expanding outer layers of the star are convective, with the material mixing through the turbulence from near the fusion regions to the surface of the star. For all stars, except those of small mass, the molten material remains in the depths of the interior of the star.

Massive stars

The most massive stars have at least three times the mass of the Sun, some have as much mass as about 50 Solar masses. The evolution of a massive star occurs in the same way as the stars of small mass until they reach the stage of the main sequence. It is in the main sequence that the star glows regularly until the hydrogen in its nucleus has turned into helium. This process takes billions of years, or even trillions of years, on a less massive star. But if it is a star with great mass, this process lasts only a few million years. After the massive stars finish melting all the hydrogen in helium, then it turns into a red supergiant, which initially consists of a helium nucleus surrounded by outer layers of expanding cold gases. Over the next million years, a series of nuclear reactions formed different elements around the iron core. The nucleus eventually collapses in less than a second, causing a large-mass explosion called a supernova, in which a shock wave expands the outer layers of the star. The supernova shines more than an entire galaxy over a period of time. Sometimes the nucleus survives the supernova explosion. If the surviving nucleus has about one-half to three solar masses, it contracts to become a tiny, dense neutron star. If the core is considerably higher than three Solar masses, it contracts until it turns into a black hole

Stellar Remaining

Once a star consumes all its fuel, its remnants can become three new stars. The factor that defines how this remnant will be, is directly linked to the amount of mass the star has had in life. These are the three types:
- White or black dwarfs

For stars of a Solar mass, the resulting white dwarf is about 1/6 solar mass, compressed to approximately the volume of the Earth. White dwarfs are stable because the force of gravity is compensated for by the degenerating pressure of the star's electrons. A consequence of the Pauli Exclusion Principle: the degeneracy pressure of the star's electrons a threshold provides for compensation for additional compression, so for a given chemical composition, dwarfs of greater mass have a smaller volume. With no more fuel to burn, the star radiates its residual heat for billions of years.

A white dwarf is very hot when it forms, with more than 100,000 K on the surface and even hotter inside. It is so hot that much of its energy is lost in the form of neutrinos in the first 10 million years and existence, and after a billion years, that star will have lost most of its energy.

The chemical composition of a dwarf white depends on its mass. A star of some Solar masses will ignite the fusion of carbon to form magnesium, neon and minor amounts of other elements, resulting in a white dwarf composed mainly of oxygen, neon and magnesium, as long as it masses enough to stay below the Limit of Chandrasekhar, and provided that the ignition of the carbon is not so violent as to explode the star in a supernova. A star with mass of the order of magnitude of the Sun will not be able to have the ignition of carbon fusion, so it will have a composition of carbon and hydrogen, with no mass enough to collapse, unless there is a later addition of matter. A star with mass less than half the Sun will not be able to have the ignition of the helium melt, which will result in a white dwarf composed mainly of helium.

At the end of the star's life, all that remains is a dark, cold mass sometimes called black dwarf. However, the Universe is not old enough that a dwarf star can already exist.

If a star grows above the limit of Chandrasekhar, which is 1.4 solar mass for a star composed mainly of carbon, oxygen, neon or magnesium, the pressure of degeneration (due to electronic capture) cannot prevent the collapse of the star. If the electronic capture of the star elements and its last fusion product (carbon, hydrogen, helium ...) is easier and the core temperature is higher, it can favor nuclear fugitive reactions, which will interrupt the collapse and lead to a type supernova. These supernovae may be many times brighter than Type II supernovae, although those in the latter case have more energy release. This infeasibility of collapse causes that no dwarf more massive than a 1,4 Solar mass can exist, with a remote exception for the stars with very fast rotation, whose centrifugal force partially compensates the weight of its matter. Mass transfer in a binary system can cause an initially stable dwarf to exceed the Chandrasekhar limit. If a white dwarf forms a closed binary system with another star, the companion's hydrogen can migrate to the dwarf until it warms enough to establish a fusion reaction, although the dwarf star remains below the Chandrasekhar limit.

Neutron Star

Stars with mass greater than 10 times the mass of the Sun, have an electrifying end: the outermost layers are ejected with violence. At the same time that the nucleus is compressed by the force of gravity, this until it turns a neutron star. Neutron stars are extremely small - no larger than the size of a city - and extremely dense. These stars usually are only about 10 km in diameter and are made up almost entirely of subatomic particles, the neutrons: this is because when the star nucleus collapses, the pressure causes electronic capture, which converts the vast majority of protons into neutrons. These stars are so dense a teaspoonful of their matter would weigh about a million tons. We call neutron stars like pulsars, because they spin fast and emit two beams of radio waves that sweep the sky and are detected as very brief pulses.

Black holes

In a simplified way, a black hole is a region of space that has such a large amount of concentrated mass (thus increasing its power of gravity) that nothing can escape, nor the light. Although their own name suggests a gap, these stars are the densest objects in our Universe and the cause of their gravitational force is there: these bodies can have the mass of stars ten times the mass of the Sun, all concentrated in the size of the city of New York.

For the birth of a cosmic object of this, it is necessary the collapse of star with at least 20 times the mass of the Sun. Larger stars die faster and their final collapse releases much energy. The star's core implodes and temperatures can reach up to about 55 billion degrees Celsius. In this final event of the star the atoms are shattered into electrons, protons and neutrons, which are still shattered into smaller and smaller pieces. This process of fragmentation continues until the star becomes a black hole. This body becomes a cosmic drain, and the speed necessary to escape from there would have to exceed the speed of light. It would be impossible for us to admire what happens beyond the horizon of events that is the dividing line between the inside and the outside of the black hole.

Between the years 1900 and 1950, the idea among physicists that an object became so dense as to draw the light to itself was disposable, although as early as 1783 the English astronomer John Michel mentions the idea and sends a report on to the Royal Society of London. Several names were given to this object: frozen stars, dark stars, crumbling stars or also Schwarzschild's Singularity. It was only in 1967, in a lecture for

Columbia University in New York, that physicist John Wheeler mentioned the term black hole.

Albert Einstein, who created one of the means to understand the cosmos, which was one of the most brilliant thinkers in physics, was never convinced of the existence of these bodies, although his equations showed this to him. In his mind it was impossible for nature to permit the existence of such objects. The idea that gravity could impose itself on electromagnetic and nuclear forces by causing the nucleus of an enormous star to simply disappear to Einstein was so unusual that it could never exist.

Increasing studies show that most galactic nuclei. Just a bustling concentration of stars, gas and dust. On the axis of this chaotic agglomeration, and in practically all galaxies including our Milky Way, there is at least one of these supremacies and compound objects, relying on a tremendous gravitational force. Whether it is speculated, manipulated mathematically or not, it has become a fact and there are about trillions of them out there. Black holes are very famous though many things about it still are fruit of science fiction stories: the first is on the question of a black hole sucking people. The black hole itself does not come out sucking all around, it only has an extraordinary attraction to its size. I have the particular opinion that when we talk about black holes, the best comparison is not with a cosmic drain, but rather, in a better analogy, it would be with a time machine. Surely you heard the phrase that says time is relative. Time is a difficult concept to understand, but Einstein understood. He understood that time is affected by gravity: to make this clear, it is interesting to mention about Global Positioning Satellites (GPS) clocks, they are programmed to work slower than Earth clocks: if this were not the case, GPS would not be correct. When we are close to the event horizon, what all fiction writers dream of becomes reality: gravity overcomes time and every minute there is equivalent to a thousand years here on Earth.

The black holes have not always been that massive and gigantic, but with every cosmic meal they get more and more fat. It is very probable that they will be the last stars of our Universe: after devouring everything that makes up the cosmos, they will be the last celestial objects and this will probably happen in about 10 thousand trillion years. All fat, full, with mass of at least 1 billion times the mass of our Sun; This after devouring all the light of our Universe. And when there are only so much black holes, the pitch will be eternal and no one will be alive to turn on the light.

Southern Hemisphere Stars

When you look at the southern hemisphere sky, you are looking toward the galactic center, which has a huge population of stars. As a result, the Milky Way appears brighter in the southern sky than in the northern sky. The southern sky is rich in nebulae and star clusters. It contains the Great and Small Magellanic Cloud, which are the two closest galaxies. The stars form fixed sets in the sky, called constellations. However, the constellations are only apparent groupings of stars, since the distances between the stars within the constellation can vary enormously. The shapes of the constellations can change over many thousands of years, due to the relative movements of the stars. The movement of the constellations in the sky is caused by the movement of the Earth in space. The Earth's daily rotation produces the movement of the constellations skyward from east to west, and the Earth's orbit around the Sun makes different areas of the sky visible at different seasons. The visibility of the sky areas also depends on the location of the observer. For example, stars closest to the celestial equator can be seen from both hemispheres during a certain period of the year, while stars near the celestial poles can never be seen from the opposite hemisphere.

Northern Hemisphere Stars

When you look at the northern hemisphere sky, you are looking out from the more densely populated galactic center. Hence, the northern sky generally appears less brilliant than the southern sky. Among the most well-known aspects in the northern sky are the constellations Ursa Major and Orion. Some ancient civilizations believed that the stars were fixed in a celestial sphere that surrounded the Earth, and the modern maps of the sky are based on a similar idea. The north and south poles of this imaginary celestial sphere are situated directly above the north and south poles of the Earth, points at which the axis of earth rotation intercepts the sphere. Polaris is very close to the center of the north pole. The celestial equator delimits the projection of the terrestrial equator on the sphere. The ecliptic marks the trajectory of the Sun through the sky, while the Earth orbits around the Sun. The Moon and the planets move through the bottom of the stars, because the stars are farther apart; the closest star to the Solar System (near the Centaur) is 50,000 times as far away as the planet Jupiter.

The constellations

The orientation was one of the necessities that motivated the study of the sky, since the man stopped being nomadic, now had a fixed place to where it returned after the period of hunting. At that time, it was necessary to gather the stars in groups to facilitate their recognition, for it gave the name of some objects, animals and gods; so, the constellations were born. Another very important factor for the creation of the constellations was for the ancient peoples to be able to watch the seasons. Some historians believe that much of the myths associated with the constellations were created so that former farmers knew the time to plant and harvest.

Most of the ancient cultures saw images in the clusters of stars in the sky. Early efforts to catalog the stars can be seen in ancient cuneiform records dating from 6000 BC. Found in the valley of the river Euphrates (Babylonians). Leo, Taurus, and the Scorpion were already seen by the ancients in the sky. The constellations of today are, however, different from those of yesterday, as they are the contribution of various societies and peoples. Greeks and Romans were, however, the ones who contributed most to the formation of myths and figures in the sky.

In 1922 the International Astronomical Union created a catalog with 88 such constellations as we know today, officially determining that all constellations be classified and listed in alphabetical order with their Latin names. Most of them had been known for a long time, 42 of them were cataloged by Ptolemy in his book called Almagest, in the second century; the others were defined in the seventeenth and eighteenth centuries. The most recent constellations were defined by Nicolas Louis de Lacaille in his book called Caelum Australe Stelliferum, 1763.

Today, in modern astronomy, a constellation is an internationally defined area of the celestial sphere. These areas are grouped around asterisms, which are patterns formed by stars, which are apparently close to each other. But there are numerous historical constellations that are not recognized by the UAI, as well as constellations recognized in regional traditions of astronomy and astrology, such as Chinese, Hindu or Aboriginal Australian.

We will speak of the constellations seen by every ancient people who lived and left their mark in the heavens and records to tell us this. In essence, the purpose of the constellations had a character attached to the beliefs of each people, reaffirming myths, stories, omens, divine signs, and so on. Moreover, the constellations also served as a map, now called a celestial chart, and in this case, they also served for the study of astrology, as an astral map, although in the case of astrology, there is more emphasis on the constellations of zodiac, on the planets and on the moon.

A small Latin class

As the International Astronomical Union has defined constellations to be classified and listed in Latin, most of the names are written in this language. The stars of the constellations are given Greek designation. So, when you read Alpha Canis Majoris, we are talking about the alpha star of the Great Dog constellation, in this case, Sirius. To form this name, we use the designation of the star (alpha, beta, gamma ...) with the genitive of the constellation.

Aries (Aries)

-Located in the equatorial zone-

It is a very old constellation, initially referred to as "the peasant." Subsequently, the ancient Babylonians, Egyptians, Persians and Greeks, called this group of stars the Ram.
A plague was devastating the region of Thebes, where Atamas reigned. A plague was devastating that kingdom, and Ino, who hated the king's sons (of whom she was a stepmother), invented that the oracle replied that it would only be possible to pass that pest if the king's sons were slain. When they were about to be sacrificed, Hermes sent a ram with the golden coat and took the children. The sons of King Frixo and Hele would be taken from Europe to Asia by the Golden Ram. However, as Hele passed the strait separating Thrace from Troas, he was startled by the sound of the waves and fell into the sea. For that reason, that part of the sea received the name of Hellespont, or Sea of Hele. Frixo tried to rescue his sister, but in vain. Then he continued his journey and as soon as he landed, Frixo offered Zeus the ram that saved his life. Immolated the animal and hung it in a forest consecrated to Ares. The gold coat of the animal was being guarded by a dragon that should devour anyone who approached. The ball became known as the Golden Fleece and the Powerful Zeus, who was very satisfied with the sacrifice, said that it would

guarantee happiness and abundance to those who approached that ram. The problem was getting that done.

Andromeda, Cassiopeia, Cetus and Cepheus

-Located in the northern hemisphere-

It is one of the earliest constellations to be baptized, and its antiquity gave it time to generate a rich and varied mythology around it, since it incorporates the legends of other groups of stars identified later.

Cepheus and the beautiful Cassiopeia were the ruling kings in Ethiopia and it was there that Perseus decided to go to rest after killing Medusa. They had a daughter, even more beautiful than her mother, Andromeda. Both boasted of being more beautiful than the Nereids, and even more beautiful than Hera, the queen of Olympus. they became angry with the feat and complained to the god of the seas, the mighty Poseidon. The latter, outraged, flooded the entire verdant plain of Ethiopia, and the population began to starve and miserable. And since it was not enough punishment, the population began to be haunted by a sea monster. King Cepheus went to find the oracle who told him: "The only way to atone for the pride of your wife and daughter, and thus appease the divine wrath, will be to chain Andromeda into a rock, so that the monster will carry out the punishment. , who was returning from an adventure, saw the beautiful Andromeda, naked and chained to that rock, which was being whipped by the waves, and there on the horizon, hour or so the back of the seahorse presented itself. The monster surfaced waves, Perseus mounted on Pegasus and rose flying over the sea. He tried to kill the monster with the sword Hephaestus had given him, but it did not work, his armor was too hard. As they stood facing each other, Perseus remembered Medusa's head and showed the monster, who was instantly frozen. The hero Perseus married the beautiful Andromeda and thus, gave rise to the Persian People. Cassiopeia is represented in heaven sitting on a throne, combing her hair, not only represented her beauty, but also her vanity.

Aquarius

The constellation Aquarius dates from the times of Babylon and is opportunely located in the sky, not far from the constellation Dolphin, the constellation River, the sea serpent and the fish, so this region of the sky is known as Sea or Water.

There are many mythological associations for this constellation, appearing sometimes identified with Zeus, pouring the waters of the life of the heavens. but one of the most beautiful legends associated with this constellation is related to Ganymede waterhole. Legend has it that he was a very educated, gentle and handsome young shepherd who was so admired by the gods that he was given the ambrosia and the nectar of the gods to be transformed into an immortal. One day, while guarding the herd and playing with his dog, Argos, Ganymede was kidnapped by the eagle, a giant eagle of Zeus, who took him to Mount Olympus, the abode of the gods, to become the favorite waterman of all the gods. There was a day when the sensitive Ganymede asked Zeus to let him help the people of Earth: Zeus, who was not usually very generous, accepted the request of Ganymede. This realized that sending a large amount of water to the Earth at the same time could become dangerous, so he decided to send it in the form of rain. That is why the young shepherd is known as the god of rain; coincidentally (or not) the birth of the aquarium constellation marks the beginning of the waters season. King Trohos, the father of Ganymede, missed Ganymede very much, and not even all the precious gifts sent by Zeus could appease this longing. Thus, Zeus placed Ganymede in the sky so that his father could see him every night.

Aquila

-Located in the southern hemisphere-

Aquila is of very ancient origin and represents the eagle that accompanied Jupiter, the leader of the gods in

Roman mythology. In Greek legend, this constellation was associated with the eagle that belonged to Zeus.

The celestial representation is intended to illustrate the episode in which Zeus sent his eagle to kidnap Ganymede, for its beauty, although a variant of this legend says that the animal is the metamorphosed Zeus himself. According to Greek mythology this would be the eagle charged with punishing the Titan Prometheus chained to the summit of a hill for having dared to share the secret of the domain of fire with Mankind, eating the eternally renewed liver, every day, until it was slaughtered by the mythical hero Hercules. The image of the Eagle would then have been placed in heaven by Zeus, in honor of his faithful dedication.

Auriga

-Located in the northern hemisphere-

A constellation known from antiquity, this beautiful figure of multiple faces is easy to find, thanks to a very bright star, Capella.

In Greek mythology, this constellation represented Melissa and Amalthea, daughters of the king of Crete, who would have nursed little Zeus with goat's milk. That is why the medieval celestial cartographies represent Auriga as a young man with a goat on his shoulders and two children on his left arm. But there is a second interpretation. Auriga would represent Erechtheus, the son of Hephaestus (the Roman god Vulcan), who invented a cart to move his crippled body.

Capella has been regarded as the goat star since the time of the Romans. Almost fifty lightyears away, it looks like our Sun, but it's bigger.

Canis Major

-Located in the equatorial zone-

The Greater Dog constellation is perhaps one of the most striking of the celestial sphere, and one of the protagonists is the brilliant star Sirius, also known as the Dog Star, the third brightest element in our sky.

The term "Canticle", very archaic, comes from a time when Sirius was high in the sky during the summer. This was invisible during this season, but it was believed that its heat joined with the heat of the Sun, thus creating an unbearable heat and for that reason was called doggy, or day of dog. Nowadays, when we go through a very difficult day, we also use that expression.

The constellation of Big Dog and its neighbor Small Dog, appear in many legends. In one, the two dogs were patiently sitting under a table where the Twins were dining. The weak stars that we can see scattered across the sky, from Small Dog to Gemini, would be according to the legend interpreted as crumbs that they gave to the animals. According to Greek mythology, the Big Dog was an animal that ran at frightening speeds. Laelaps, as they called him, had even won a race for a fox that was said to be the fastest in the world. Zeus would have placed this dog in the sky to commemorate his victory.   Another mythological legend presents the two dogs (Big Dog and Small Dog) as helpers to Orion the hunter, helping in his favorite sport that was precisely the hunting. With his gaze directed at Lepus and crouched beneath Orion, the Big Dog seems willing to leap to grab his prey.

Canis Minor

-Located in the equatorial zone-

The constellation of Small Dog is the smaller companion of Big Dog. It presents only two stars of luminosity superior to 5 magnitudes: Procyon (that in Greek means "before the dog") and Gomeisa.

It appears often referenced as one of the hunting dogs of the Orion, and shares some legends with its companion (to see constellation Dog Greater).

It is also said that Little Dog was one of the Acteon's dog. One day, Acteon surprised Artemisia, goddess of hunting and woods, while she bathed with her companions. Fascinated by her great beauty, Acteon paused for a moment, eventually being seen by her. Furious, for being seen naked by a mortal, Artemisia turned him into a deer and threw his pack of dogs at him, with Acteon being devoured.

Capricornus

-Located in the equatorial zone-

The term Capricornus means "the one that has goat horns". This is a constellation of the zodiac and is the first: the sun passes there between January 19 and February 15, and this is a very old constellation. Its stars have weak glare and for that reason is not very easy to identify it in the sky. Greek mythology says that this constellation represents the god Pan, who resembled a goat. Pan was very hesitant, he could never make a decision quickly. According to Greek legend, once the gods ran in flight from a sea monster called Typhon and to mislead the monster they turned into animals. It turns out that Pan was in doubt of which animal would be transformed, and when it saw the shadow of the marine monster approaches, without being able to decide in which was going to transform, made his trunk turn goat and his legs in fish tail: turned into goatfish.

For thousands of years the sun has reached its most southern position in the sky (when the winter solstice occurs) in front of the constellation Capricorn. During this time the Tropic of Capricorn was in South latitude.

Yet it is called, though the sun, as a result of precession, is now in the constellation of Sagittarius during the winter solstice.

Capricorn is the least conspicuous constellation of the Zodiac, and is uniting the three brightest stars of the Eagle constellation, in a line drawn south.

Scorpius

- Located in the equatorial-

In Greek mythology, Scorpius is the scorpion that killed Orion. The scorpion was the animal sent by the goddess Gaia to confront the hunter Orion, his own son who boasted of being able to defeat any animal of the face of the Earth. However, the giant Scorpion managed to sting Orion, causing his death. According to this legend, the hunter would have been resurrected by Asclepius (represented in the sky by the constellation Ophiuchus (The serpentary), due to his unusual knowledge of medicine.

Associating the legend with the celestial representation of these personages, the Scorpio finds itself, in the sky, under the foot of Asclepius, as if to be crushed or imprisoned by him. In addition, Orion and the Scorpion meet at opposite points in the sky, so that when one of these constellations "is born", the other "sets", as if both continued eternally a mutual and impossible persecution.

It is a beautiful constellation of the Zodiac, full of bright stars and rich starfield.

The pattern is unmistakable: the hideous animal in the sky, "walking" slowly at night, seems to overcome the condition of mere constellation, suggesting a real entity sneaking among the stars, with their tongs and stings ready to hit some victim.

Gemini and Cygnus
-Located in the equatorial zone-

According to legend. Zeus fell in love with Leda, wife of the king of Sparta, Tindoro. Zeus.

To approach his beloved, he became a beautiful swan (which is also represented as a constellation). From this passion were generated two the two twins Castor and Polideuces (Pollux is the Latin version). The two had the best teachings and Castor became a true gentleman, while Pollux became a brave warrior.

Once they challenged two other young men to a duel, fighting for the hand of two young women already committed. In this challenge, Castor was eventually killed. In desperation Pollux tried to kill himself, but he was immortal, so he went to Zeus and asked him to kill him too. Zeus was moved by the story and turned them into a constellation. Thus, the two young men appear in the embraced sky. There is a mystical current that gives the constellation a richer symbolism: the two boys would be, in fact, Apollo, brightness and light, and Hercules, strength and courage. Perhaps that is why in many treatises one of the twins appears holding bow, arrow and lyre, while the other appears with a club.

Heracles, the Milky Way, Hydra, Leo, Cancer and Draco

-Located in the northern hemisphere-

Heracles was the son of Alcmena and Zeus, and as soon as he was born, Hera sent two serpents to attack him in the cradle, but he strangled them both. Zeus, in order to make him immortal, took the boy and when the beautiful Hera was sleeping, put him to suck the breasts of the supreme goddess. As soon as she woke up and saw that child there, on her breast, she pushed him to get free, and the milk that flowed remained in the sky being known as the milk path. Heracles was created by a prince named Eristeu, who was afraid of being dethroned by the hero and always kept him busy, so Eristeu gave him a lot of chores that became known as the 12 works of Heracles.

One of this 12 work happens when Heracles is going to kill the Hydra of Lerna marsh, a serpent of several heads and each time you cut one, two others were born into place. While fighting with the monster, there was a crab to hinder him, poking his feet, and one trampled, crushed him Heracles. The crab was transformed into a constellation by Hera, who was the enemy of Heracles.

Another constellation associated with this is the Lion, which represents the Lion of Nemean, and Heracles' first work was to kill this terrible beast. As no arrow was capable of killing him, the hero went into bodily strife with him and strangled him. Heracles removed his skin and turned it into his shield, since it was impenetrable.

Draco, another constellation associated with this, was the dragon that tended the Garden of the Hesperides, where the golden knobs were. The 11th work of Heracles was to take one of these knobs. It turns out that this garden was guarded by a huge dragon, which Heracles killed.

According to legend, a centaur intended to kidnap his wife, and Heracles kills him with an arrow poisoned with the blood of the Hydra from the swamp of Lerna. Before he died, the centaur gave the woman his blood-stained clothes, telling him to persuade Heracles to wear them, for they would avoid any kind of infidelity of her husband. However, these were contaminated with poison, and as soon as the hero wore those robes, he felt the poisoned blood running through his veins. If he tried to take it off, he would rip his flesh too. Unable to bear that pain, the hero ascended Mount eta, where his great friend Philoctetes lived. Hercules plucked the pines out of the forest with his own hands, making a huge pyre, and asked his friend to light the fire, which, weeping heavily, answered his friend's request, putting an end to his pain. But what burned was only the mortal part, the hero went up to Mount Olympus, the abode of the gods.

Orion

-Located in the southern hemisphere-

Orion was the greatest of the giants, son of Poseidon, the god of the sea. Legend has it that he was also the son of Gaia, Mother Earth. He was a powerful giant. From the father, he had inherited the power to walk on the waves of the sea. From her mother, she had inherited the gigantic bearing. With the times, he was beautiful and athletic, with a beautiful figure, coveted by women and goddesses. He married the first wedding party, which was said to be the most beautiful of all the young women of ancient Greece. But Side was proud and boasted of being even more beautiful than the immortals, more beautiful than Hera herself, the queen of the goddesses and wife of Zeus. Hera took her vengeance and hastened the girl from the top of the Tartarian mountains, killing her.

Deprived of his wife, Orion pretended to be lost by the Earth. At one point, as he passed by the island of Chios, famed for his thick hunting, he saw Merope, the princess of the kingdom, who played his flute on the banks of a river. As soon as they saw each other, the young men fell in love. Merope had never seen such a handsome giant, and Orion had never seen a young woman so innocent and so gifted. But their loves were also destined for tragedy. Her father, King Enopion, was known to have introduced red wine. Its name, in Greek, means "what drinks wine". At that time, the wine was still little known and Enopion managed to deceive the passers-by and intoxicate them. He was against the love of the two young people, managed to intoxicate Orion. When the giant was fucked in the heavy sleep of alcohol, Enopion blinded him with a sword and managed to drive him out of the kingdom. However, Orion got the help of a boy who sat on his shoulders and guided him, managed to walk to the Rising Sun.

When the dawn goddess saw him, she fell in love with the young giant and decided to help him. With his arts, he was able to recover his sight. Orion spent some time with the goddess, but his loves were short and soon set out for new conquests. Throughout his travels, Orion had become an accomplished hunter. With his long legs and agility, he was proud that there was no animal to escape him. With his sword and club, a club made of the tallest pine he had found, he was proud to be able to kill any animal that existed on earth. Orion spent days and days on the hunt, which became his passion. In the course of his wanderings, he would meet the goddess Artemis, whom the Romans called Diana, and who was also famous for passionately hunting wild animals.

The two famous hunters of the ancient world came together and created a strong friendship. During the day, they set off in search of new forests where they found wild animals. At night, they sat around the campfire and told each other their adventures. Legend has it that they were in love, but the truth is that Artemis, always young and chaste, thought only of sports, the outdoors and hunting, she was a mysterious character. They say she was the moon goddess, just as her brother Apollo was the sun god, but in fact she only wandered on full moon nights; also said to be cold and vindictive: it killed by taste and the truth is that Apollo and the goddess had often entertained to massacre young warriors. Perhaps their taste for blood had stayed them from the ancient combats with the giants who challenged Zeus, the chief of the gods that the two brothers helped in this ancient war to dethrone the Titans. Artemis and Orion were friends, but the goddess, who always remained a virgin, wanted nothing more than friendship. What happened to Orion, nobody ever knew. It is possible that he had fallen in love with Artemis, who was very beautiful, with athletic beauty that should have pleased the giant hunter.

What happened next no one could explain. One day, as the giant wandered through the lands of Delos, a gigantic scorpion appeared ahead of him. Orion was used to crushing these creatures. But this scorpion was larger than any of the animals that existed on Earth. He was bigger than the young hunter, and he had a shell that even Orion's sword could not penetrate.

Some say that the goddess Artemis sent the scorpion herself, for one-night Orion, not resisting desire, had wanted to rape her. Others say that Gaia sent the beast herself, Mother Earth, for Orion had boasted that there was no animal that the Earth believed he could not overcome. What is certain is that a furious struggle ensued, and that the gigantic scorpion, impenetrable to the hunter's sword and indifferent to its blows, was able to strike him with the poisonous sting of his Shaula. Orion was already dead, and still the scorpion continued to smite him when Zeus appeared. Impressed by the power of the beast, the chief of the gods took him to the heavens. Commuted with the heroism of the vanquished giant, he also carried him to the sky, but placed him in the opposite position to his victor, so that the two enemies could be in the heavens without ever seeing each other. So, it is today: when spring begins, Orion disappears in the sunshine; when autumn appears, the dangerous scorpion is swallowed by the horizon of sunset. And when we observe the sky, it always seems that one is hunting for the other.

Sagittarius

-Located in the equatorial zone-

It is one of the twelve constellations of the zodiac and one of the oldest in the sky. Legend has it that Sagittarius is a centaur half man, half horse and is identified with Chiron. He was the son of Kronos and Filira. A centaur different from the rest of his kind because of his great wisdom. His father taught him medicine, astronomy, and music. Chiron was charged with the education of several princes and heroes, such as Achilles, Theseus, Ulysses, and Jason.

But when beside Heracles, he fought against the centaurs, he was accidentally hit by an arrow shot by the hero. Suffering terrible pain but prevented from dying, Chiron yielded his immortality to Prometheus and Zeus placed him in the Zodiac, where he appears as the constellation of Sagittarius. Legend has it that Chiron created the constellation to guide Jason and the Argonauts when they sailed on the Argos.

Ursa Major and Ursa Minor
-Located in the northern hemisphere-

The observation of the Ursa Major constellation occurs more easily between January and October, when it is not so close to the horizon, depending on the location you are on the planet. This constellation can be called Car, Plow and even Frying Pan or Casserole. The Big Dipper is a constellation easily recognized by its seven bright stars that draw a square and a tail. This figure is an asterism, that is, a characteristic grouping that does not constitute a constellation, since it is much larger (it extends over a large area of the sky and includes about two hundred other stars visible to the naked eye). The stars of this constellation, being near the pole, are always visible on nights of almost the entire northern hemisphere, so it is also known as the circumpolar constellation.

According to Greek mythology, Zeus fell in love with Calisto, the beautiful nymph of the woods and companion of Artemis. Zeus was so fascinated by her beauty that, to approach her, he took the features of Artemis. Calisto welcomed Zeus without suspicion, but when he recognized his error it was already too late, and he conceived of him a son, who was called Arcas. Hera, wife of Zeus, was furious and chastised Calisto. By irony of fate, Calisto, who loved to hunt, turned the hunt: Hera had turned her into Ursa. One day she and her son, Arcas met, and when Calisto opened his arms to welcome him, he thought he would be attacked by the gigantic bear, and prepared to kill her. At the last minute, Zeus avoided the tragedy and turned Arcas into a small bear, dragging them both to the skies. Hera, however, pushed them both close to the north pole where the stars are always visible mother and son would never have rest, and Arcturo, stood guard to the bears so that they did not move away from the icy pole.

Boötes

-Located in the northern hemisphere-

The word Boötes is of Greek origin and means drover. The constellation represents the son of Demeter, who according to the legend awarded with a place in the sky for having invented the plow. According to legend, Drover was a very sensible young man with a great sense of dedication and social conscience. When faced with the difficulty in which the inhabitants of the planet Earth found themselves to obtain food, Drover decided to help them to be able to produce their own food, created the plow and sent it to Earth. It was because of this help to humans that the gods, the gods decided to reward him by giving him the honor of being placed in the sky, at the position of the Big Dipper (also known as plow). Another legend transforms Boötes, also known as Arcade and Arcturus, into the son of Zeus and Calisto. This one, transformed into bear by Hera, the jealous woman of Zeus, almost was assassinated by its son when this one was hunting. Zeus rescued her and led her to the sky, where she became the Big Dipper. The name Arcturus (the brightest star in the constellation) comes from the Greek and means "bear guardian".

Next to this constellation is located another constellation known as Canis Venatici, although the stars constituting it are already known by the Greek people, were an asterism that was part of the figure of the constellation of Drover, in the form of a stick that was bellowed by the shepherd. Thus, this asterism was not seen as an independent constellation, so there is no mythological legend associated with it.

Apparently, the Greek word "bat" came to appear later, for translation errors, as "dogs."

It was Hevelius, a Polish astronomer who, trying to identify these dogs that were referred to in the Latin translation of Ptolemy's Arabic version of Almagest, concluded that these would be the stars that would represent them.

Having been the first to publish in a celestial atlas, strict with what we observe in the sky, the figures of two Hound Dogs trapped by the leashes in the hand of the Drover, the Hevelius became recognized the authorship of the modern constellation Canes Venatici. The proper names of each of the dogs, *Asterion* and *Chara*, although etymologically have Greek origins, were also attributed by Hevelius: *Asterion* translates as Little Star, while *Chara* means Joy.

Perseus and Pegasus

-Located in the northern hemisphere-

Perseus is a beautiful constellation that crosses the Milky Way. Its stars form an arc from Capella, in the Coach to Cassiopeia.

According to the associated Greek legend and this constellation, Perseus was the son of Danae (daughter of Acrisius, king of Argos), and Zeus. Danae had not yet given an heir to the throne of Argos, and Acrisius was troubled, and decided to seek the oracle of Delphos to know whether he had a successor. Two answers waited for him, one good and one bad. The positive answer was that he would rather have a successor, but the negative answer was that his own grandson would take his throne and his life. Annoyed, Acrisius returned home and locked Danae in a bronze tower, so that she would be cut off from the world and never have any children. However, when an oracle predicted something, it never failed. Zeus fell in love with the beautiful Danae and turned to rain of gold, and made a child in the beautiful virgin, Perseus. When the boy was born and old Acrisius discovered, he threw them both into the sea, the winds blew and the coffin they were sealed reached the island of Zerfiro, where King Polidectes received them. This king fell in love with the beautiful girl, but that child disrupted his plans. Polidectes would wait until Perseus reached adolescence and to have the free way with the beautiful Danae, ordered the Young to fetch the head of Medusa, the goddess who turned to stone whoever looked at her head. He fought the Gorgons and cut off the head of the goddess, as soon as he cut off the goddess's head, a beautiful white winged horse was born from the blood, which became known as Pegasus. On his return journey, he encountered Andromeda (see story above), Princess the hero saved and married. At the end of the honeymoon, he returned to the Island of Zefiro and discovered that his mother had to flee, as she was being mistreated by her stepfather. He went to Polidectes, who insulted him by saying that he had not killed Medusa. At this moment, Perseus removes the head of the Medusa from the bag, transforming the king Polidectes into stone.

And the oracle? He never failed, would not fail this time. When Perseus, Danae, and Andromeda were traveling, they passed through Argos. At this very time the Olympic Games were held, and Perseus offered to play records, but when playing, he misses and hits someone's head in the audience: King Acrisius, his father.

Corvus, Hydra and Crater
Both of the South-

These three constellations in keep one close to the other. According to legend, Apollo had ordered his pet bird to fetch water from a distant spring, in a Cup (also represented by a Crater constellation). However, the crow found a fig tree on the way and could not resist eating the figs, so it landed and waited for a few days for the fruits to mature. When he came to the spring, when he saw a snake of water, I remembered to make an excuse to justify his delay: he told Apollo that the presence of the snake had made him wait to collect the water. The power to see through the lie was characteristic of this Greek god by which, as punishment, he placed the Raven, the Hydra (the water snake) and the Cup in the sky, giving orders to the snake to never let the raven reach the water, in order to feel thirst for eternity.

Draco

There are some ancient legends related to this character, being associated with several different dragons present in Greek myths, sometimes identifying even the exact same monster in different contexts; these contradictions are explained by the fact that these legends originated in distinct classical authors who, independently, dedicated themselves to explaining the presence of the character in the sky.

The best known of all is the one that is said to be the dragon that kept the golden apples of the Garden of the Hesperides, in one of the episodes of the 12 works of Hercules. The mythical hero was able to kill (or, in other versions, fall asleep) the dragon and reap the apples, having the goddess Hera then placed the image of the monster in the sky.

Another Greek legend tells that this would be the Dragon of Colchis, killed by Jason so that the mythical hero could obtain the Golden Fleece. A lesser-known myth, among many others, says that this would have been the monster killed by another mythical Greek hero, Cadmus, founder of the city of Thebes, to reach a fountain in a sacred grove guarded by a dragon.

Taurus

From the antiquity to the present, the Bulls have been objects of worship and worship as symbols of strength and fertility. Their presence in legends and representations is constant. In classical times, the Greeks believed that the constellation of Zeus was a Taurus in disguise. Legend has it that Zeus fell in love with beautiful Europe, the daughter of Agenor, king of Phenicia. He was enchanted by her beauty and turned into a white Taurus, who knelt at his feet. Europe then ascended to its back and adorned its horns with flowers. Leaping to his feet, the Taurus headed for the sea, swimming to the island of Crete, where Zeus made Europe his mistress. They had three children. One of them, Minos, would later become King of Crete. In the constellation you can only see the bull's forequarters, as if it were coming out of the waves.

It may also represent one of the 12 works of Heracles, for the legend says that Minos, the king of Crete, wanted to please Poseidon and promised him to sacrifice anything that God would bring forth from the waves. Then from the waves came a beautiful beast, which Minos worshiped and sacrificed another rickety ox. Poseidon, P. of life, ordered the bull to devastate the whole island. Heracles seized the animal's horns and subdued it.

Ophiuchus
- Located in the equatorial zone-

According to the most famous legend, full of variants, it represents Asclepios, son of Apollo with the mortal Coronis. Coronis commits an infidelity while pregnant with Asclepius, eventually falling victim to the vengeance of Apollo, who saves Asclepius from his mother's womb, delivering him to Chiron, leader of the Centaurs, to teach him the noblest arts. Asclepius becomes a physician of unparalleled skill, capable even of resurrecting mortals.
Unable to allow such a thing, Zeus kills him with one of his lightning bolts, provoking Apollo's wrath against the leader of the gods of Olympus. He then tries to avenge himself on Zeus, killing the Cyclopes who made his lightning bolts, and by punishment Zeus places Asclepius in the sky so that Apollo feels the pain of seeing the figure of his son and at the same time is reminded that no more you can bring it to the world of the living... and mortals. A variant of the legend says that, unlike the version presented earlier, the placement of the

Asclepius figure in the sky would seek to console Apollo in his pain.

The constellation of Ophiuchus (or Serpentary) nowadays forms part of the group of 13 modern and official constellations through which the Sun passes during the year, an apparent path called the Ecliptic. It should, from the outset, be associated with astrology to one of its Zodiacal Signs.

Virgo
-Located in the equatorial zone-

This is a very old constellation. Curiously, Virgo is the only female figure, among the constellations of the Zodiac, and has symbolized an extensive range of deities since the prolegomena of history stelae. Generally, in its representations, Virgo appears with a spike of wheat, or carrying the scales of Libra, the adjacent constellation.
One of the legends associated with this constellation relates that the Virgin was Erigone, daughter of Icarus to whom Dionysius had uncovered the secret of wine production. When they first tasted and felt the haunting effects of this drink, the local villagers believed that Icarus had tried to poison them and killed them. Icarus's dog, Maera, soon attracts the attention of Erigone to the murder of his father, who, being unable to bear such sadness, ends his life by hanging himself. The gods, moved by this tragedy, would have placed the interveners in the sky, in the form of the constellations of Virgin Ergony), Boötes (Icarus) and Dog Minor (Maera). According to the most celebrated legend, full of variants, it represents Asclepius, son of Apollo with the mortal Coronis. Coronis commits an infidelity while pregnant with Asclepius, ending up as a victim of Apollo's thirst for vengeance, which saves Asclepius from his mother's womb, giving him to Chiron, leader of the Centaurs, to teach him the noblest arts. Asclepius becomes a physician of unparalleled skill, capable even of resurrecting mortals. Unable to allow such a thing, Zeus kills

him with one of his lightning bolts, provoking Apollo's wrath against the leader of the gods of Olympus. He then tries to avenge himself on Zeus, killing the Cyclopes who made his lightning bolts, and by punishment Zeus places Asclepius in the sky so that Apollo feels the pain of seeing the figure of his son and at the same time is reminded that no more you can bring him into the world of the living ... and mortal. A variant of the legend says that, unlike the version presented earlier, the placing of the figure of Asclepius in the sky would seek to console Apollo in his pain.

The constellation of Ophiuchus (or Serpentary) nowadays forms part of the group of 13 modern and official constellations through which the Sun passes during the year, an apparent path called the Ecliptic. It should, from the outset, be associated with astrology to one of its Zodiacal Signs.

Argos Navis
- located in the southern hemisphere,

This constellation, along with Carina (the keel), Puppis (the Compass) and Popa, was part of a huge group of stars of the southern sky, which was known as Argo Navis, Boat Argo, which was the boat that Jason and the Argonauts sailed in search of the Golden Fleece.

Only the rear half of this ship can be visualized, and legend has it that the rest of the vessel is not visible because it is shrouded in fog, or else because the episode celebrated in this constellation refers to the moment when the ship launches across the Blue rocks that would therefore be hiding the bow. Argo Navis was divided by Nicolas Louis de Lacaille around 1750, and shares its stars with the four constellations that resulted from it. This caused Vela not to have Alpha or Beta stars.

It is located relatively easily because it is made up of stars of reasonable brightness, using as reference the proximity of the two brightest stars of the sky: Sirius (of the neighboring constellation, Big Dog) and Canopus (of the neighboring constellation, Keel).

Lupus and Ara

- Both located in the southern hemisphere,

The association with a wolf appears later in the late middle ages. According to the most famous legend, originated in Greek mythology, the wolf represents an animal hunted by the Centaur, who prepares to offer it to the gods in the Altar (near constellation that represents a sacrificial altar).

In Greek mythology, it represented the altar on which the gods of Olympus exchanged vows of union before the battle against the Titans. According to this legend, Titan Cronus had deposed his progenitor, Uranus, and reigned over the Universe together with his brothers. So that none of his children would do the same, Cronos swallowed him at birth but his wife, Reia, no longer taking this situation, hid Zeus so he would not suffer such a fate. As an adult, Zeus confronted his father, making him vomit all the other brothers and joining them in a battle against the Titans. On the altar, represented by the constellation of Ara, they swore union and fidelity and then began the war for supremacy over the Universe. Zeus was forced to free two races of Titans. With their help, the gods of Olympus were finally able to defeat the Titans, so Zeus would have placed the Altar in the sky in eternal gratitude to his allies.

Lira

-Located in the northern hemisphere-

This is a beautiful constellation dominated by Vega, one of the brightest stars in the sky. Looking closely, you can almost imagine the ropes of the Lyre extended by the parallelogram of four stars that accompanies it.

Its origin is very ancient, being also present in the Greek culture, that in it he saw the representation of the musical instrument invented by the god Hermes, from the carapace of a turtle. The Lyre was played sublime by Orpheus, son of Apollo who had offered him the instrument. Orpheus played so well that wild animals were delighted.

Orpheus was hopelessly in love with his wife Eunice, and when she died, he descended to hell to save her. She asked the gods to release her, to which they acceded, provided that she did not look at her on the journey.

But Orpheus, impatient, looked at Eunice before they reached the upper world, and she was transported to Hades forever. Disconsolate, Orpheus was shattered by a group of women, after ignoring his insinuations. Zeus will then have ordered a vulture to collect the Lyre and place it in the sky, in memory of the art of Orpheus.

Eridanus River
-Located in the southern hemisphere-

Since antiquity this constellation has been compared to a river. In one of the earliest histories of Greek mythology, Eridanus was a divine river, born of the union of Ocean and Tethys, not being identified with any royal river. Eridanus was, however, associated with some famous rivers: the Nile, or the Po river (in Italy) known by the ancient Greeks as "Eridanus" or the Euphrates, among others. Another Greek legend refers that Eridanus would be the river where Cetus lived, (constellation The Whale).

Galaxies

A galaxy is a huge mass of stars, nebulae and interstellar matter. Planetary systems, star clusters and interstellar clouds. What is between these objects we call interstellar medium, gas space, dust and cosmic rays. Dark matter appears to account for about 90 percent of the mass of most galaxies. Some observational data suggest that there may be supermassive black holes in the center, if not all galaxies. They are believed to be the main drivers of active galactic nuclei - a compact region in the center of some galaxies that have a much higher brightness than normal. It is likely that the Milky Way possesses at least one of these objects. The smallest galaxies contain about 100,000 stars, while the largest galaxies may have more than 3 trillion of them. There are three main types of galaxies (which have historically been categorized according to their apparent form, usually repressed as their visual morphology): elliptical, which has an oval shape; spiral, which has spiral arms outside the central and irregular protuberance that has no definite shape and typically originated by the gravitational pull of neighboring galaxies. Sometimes the shape of a galaxy can be distorted by collision or interacting with another galaxy. These interactions between galaxies may, in the end, result in their joining, with the increasing increase of incidents of star formation leading to the Starburst galaxies. Which has spiral arms outside the central and irregular protuberance that has no definite shape and typically originated by the gravitational pull of neighboring galaxies.

There are likely to be more than 170 billion galaxies in the observable universe. Most have between 1,000 and 1,000,000 parsecs in diameter and are separated by distances of the order of millions of parsecs. The intergalactic space is filled with a fine gas with an average density of less than one atom per CM^3. Most galaxies are organized in a hierarchy of associations known as groups and clusters, which in turn form larger superclusters. On a larger scale, these associations are generally organized in filaments and walls, which are surrounded by immense voids, as we have already mentioned when speaking about dark energy.

About the name galaxy, the word derives from the Greek term Υαλαξιας (milky) or kiklos galakticos, (milky circle) because of the appearance that our galaxy has in the sky. The name is based on Greek mythology, which tells the story of Zeus who begot a son with a mortal: the little Hercules. Zeus tried to get his son. Zeus tried to make his son become immortal, put the little one to suckle in the bosom of Hera, while she slept. During clandestine breastfeeding, Hera wakes up and pushes the baby and a jet of divine milk splashes into the night sky, producing the thin strip of light known as the Milky Way.

Our galaxy: The Milky Way

When we look at the sky, depending on your region, (and of the time) we can see a dim light that extends through the night sky. Its light comes from the stars and nebula in our galaxy, known as the Milky Way. Our galaxy is spiral-shaped, with a dense central bulge surrounded by four spiral arms outwards, contained in a larger, less dense halo. Our sun is located on one of the arms, the arm of Orion. The central bulge of the galaxy is relatively small, dense and spherical; containing older, red and yellow stars. The spiral arms, unlike the galactic center, contain warm, blue, and young stars. Our galaxy is more or less 100,000 light-years across.

Around the year 410 BC the Greek philosopher Democritus of Abdeca proposed that the bright strip in the night sky should be made up of distant stars, although at that time it was believed that the Milky Way was the cause of the ignition of the burning exhalation of some large stars, numerous and close, and that ignition occurred in the upper part of the atmosphere. Another Greek philosopher called Olimpiadoro the Young (495-570 BC) criticized this view and argued that if the Milky Way were to sublunar it should look different, at different times and places on the Earth and would have parallax, which it does not has. In his view, the Milky Way was heavenly and this idea became influential in the Islamic world. The Arab astronomer Alhazen (965 - 1037) was the first astronomer to try to measure the parallax of the Milky Way, and determined that since it was devoid of parallax, it was far from Earth.

It was the Persian astronomer Abdul Rayham al-Biruni (973-1048) who proposed that the Milky Way was a collection of countless fragments with the character of turbid stars. In the year 1610 came the confirmation that: The Milky Way consisted of many stars, it was when Galileo Galilei observed it with his telescope and discovered its stars. In the year 1750, Thomas Whight, in his work called An Original Theory or New Hypothesis on the Universe, speculated (correctly) that the galaxy should be a rotating body, composed of a large number of stars held together by gravitational forces, similar to what occurs in the solar system, but on a much larger scale.

Stellar clusters

Clusters are groups of stars of which two types are defined: gobbler clusters are concentrated groups of hundreds or thousands of very old stars that are gravitationally interconnected, while open clusters are more scattered groups of stars, usually very young. Open clusters are torn away over time by the gravitational influence of giant molecular clouds (NMGs) as they move through the galaxy, but the cluster members continue to move about in the same direction, even though they are not more gravitationally bound; They are then known as stellar associations and sometimes moving groups.

Some examples of stellar clusters visible to the naked eye are the Pleiades, Hyades and Crèche.

Government Agglomerates

Gobble clusters are groupings with spherical visual morphology of 10,000 to several million stars, concentrated in regions from 10 to 30 light-years in diameter. They usually consist of very old stars, mostly yellow or red.

Our galaxy contains about 150 gobbler clusters, some of which may have been captured from smaller galaxies, such as the M79 clump in Lepus. Some galaxies are much richer in open clusters.

Some of the brightest star clusters are visible to the naked eye, and the brightest,
Omega Centauri, has been known since antiquity and has been cataloged as a star. In the North, the best-known globular cluster is M13, called the Great Agglomerate of Hercules.

Open clusters

These clusters are very different from the gob clusters. They are usually young objects, up to a few tens of millions of years, with rare exceptions of a few billion years, for example the object cataloged by Charles Messier known as Messier 67

Opened clusters contain up to a hundred members, within a region of about 300 light-years across. Being much less densely populated than gobbled clusters, they have a much smaller gravitational link and over time, they are broken by the gravity of Giant Molecular Clouds or the gravitational force of other clusters.

The most outstanding open clusters are the Pleiades and the Hyades in Taurus. The open clusters are dominated by young, warm blue stars, because although these stars are short-lived, using astronomical terms, these clusters tend to disperse before these stars become extinct.

The intermediate forms

In 2005, astronomers discovered a new type of star cluster in the Andromeda galaxy, which at many points is very similar to the gobbler clusters, although they are less dense and have a weaker gravitational bond. So far, no intermediate clusters (which may also be called extended clusters) have been discovered in the Milky Way). The three extended clusters discovered in the Andromeda Galaxy were christened M31WFSC1, M31WF5C2 and M31WF5C5.

These newly discovered stellar clusters contain hundreds of thousands of stars, a number similar to what can be found in gobbler clusters. They may also share other characteristics with the gob clusters, such as stellar population and metallicity. What distinguishes them from each other is that these newly discovered clusters have a diameter of a hundred light years greater and are a hundred times less dense. The distances between the stars are therefore much greater in the extended clusters. Parametrically, these clusters lie somewhere between a globular cluster (little dark matter) and a spherical dwarf galaxy (with dominant dark matter). It is not yet known how these clusters are formed, but the process may be similar to the formation of other types of star clusters. Likewise, it is still unknown why M31 (another name for which the Andromeda Galaxy is known) has this type of celestial object while our Milky Way does not have. Another curious fact is that other objects such as these have not yet been found in other galaxies, but scientists believe that it is very unlikely that M31 is the only galaxy with intermediate clusters.

Nebula

The nebulae are clouds of dust, hydrogen, helium and plasma. Originally, nebula was the name of any diffuse celestial, including galaxies beyond the Milky Way. The Andromeda Galaxy, for example, was originally from the Andromeda Nebula (and the spiral galaxies were designated as "spiral nebulae") before the true nature of the galaxies was known to have been confirmed, which occurred in the early twentieth century. Vesto Melvin Slipher, Edwin Hubble and others.

There is a very large variation in the size of the nebulae, ranging from a few million kilometers to hundreds of light years in diameter. Although denser than the space that approaches them, most of the nebulae are much less dense than any vacuum created in the earth's environment - a cloud x the size of the Earth would weigh only a few kilograms.

Nebulae are often regions of stellar formations, such as the Eagle Nebula. This is portrayed in one of NASA's most famous images, the "Pillars of Creation". In these regions the formation of gas, dust, and other materials are piled up to form larger masses, in which more masses, and eventually become massive enough to become stars. The remaining materials are believed to form planets, and other objects of planetary systems.

Observational history

Around the year 150, Claudius Ptolemy recorded, in books VII-VIII of his Almagest collection, five stars that appeared to have no definite shape, nebulae. He also noticed a region of cloudiness between the constellations of Ursa Major and Leo that were not associated with any stars. The first true nebula, far from a stellar cluster, was mentioned by the Persian / Muslin astronomer Abd al-Rahman al-Sufi in his Book of Fixed Stars (964). He noted "a small cloud" where the Andromeda Galaxy is located. He also cataloged the stellar cluster Omicron Velorum as a "hazy star" and other nebulous objects, such as the Al Sufi Agglomerate. The supernova that created the Crab Nebula, SN 1054, was observed by Arab and Chinese astronomers in 1054.

On November 26, 1610, the astronomer Nicolas-Claude Fabri de Peiresc discovered the Orion Nebula using a telescope. This nebula was also observed by Johann Baptist Cysat in 1618. However, the first detailed study of the Orion Nebula was not written until 1659 by Christiaan Huygens, who believed he was the first person to discover this cloudiness.

In 1715 Edmond Halley published a list of six nebulae. This number grew steadily throughout the century, with Jean-Philippe de Chéseaux compiling a list of 20 nebulae (including eight previously known) in 1746. From 1753 Nicolas Louis de Lacaille cataloged 42 nebulae from the Cape of Good Hope, most of which were previously unknown. Charles Messier then compiled a catalog of 103 "nebulae" (now called Messier's Objects, which included what are now known to be galaxies) in 1781, his main interest was to detect comets.

The number of nebulae expand exponentially thanks to the efforts of William Herschel and his sister Caroline Herschel. His book Catalog of Thousand New Nebulas and Clusters of Stars was published in 1786. A second catalog of a thousand was also published in 1789 and the third and final one, containing 510, appeared in 1802. For much of his work, William Herschel believed that these nebulae were just clumps of unresolved stars. In 1790, however, he discovered a star surrounded by cloud cover and concluded that this was a true nebula, rather than a more distant cluster.

Early in 1864, William Huggins examined the spectra of about 70 nebulae and found that about one-third of them had the emission spectrum of one gas, while the remainder showed a continuous spectrum, so it was clear that they were sent by a concentration of stars. In the year 1912 Vesto Slipher showed that the spectra of the nebula that surrounded the star Merope combined as a spectrum of the open Pleiades set, thus showing that the nebula radiates by the light of the reflected star.

Around 1922, after the Great Debate, it became clear that many "nebulae" were in fact, distant galaxies of our own.

Slipher and Hubble continued to collect the spectra of varied diffuse nebulae, and in this search, they found 29 that had emission spectrum and 33 types that presented continuous spectrum of light of the star. Around the year 1922, Edwin Hubble announced that almost all nebulae are associated with stars and in their illumination comes from the moon coming from them. The two scientists also found that emission-spectrum nebulae are almost always associated with stars with a Bl-spectra, warmer (including all main sequence stars including type O), while continuous spectral nebulae appear with more stars cold

Types of nebulae

Emission Nebula

Emission nebulae are gas clouds with high temperatures. Atoms in the cloud are energized by ultraviolet light from a nearby star and emit radiation when they decay to lower energy states (neon lights shine in much the same way). Emission nebulae are usually red, this happens because of hydrogen, the most common gas in the Universe and that there is commonly red light.

Reflection Nebula

Reflection nebulae are dust clouds that simply reflect the light of a star or nearby stars. Reflection nebulae are generally blue because the blue light is more easily spread. Emission and reflection nebulae are usually seen and sometimes called Diffuse Nebulae.

Dark nebula

There are also dark nebulae, they are clouds of gas and dust that almost completely prevent the light from passing through them, they are identified by the contrast with the sky around them that is always more starry or luminous.

Planetary nebula

Planetary nebulae were named after William Herschel because when they first saw the telescope they looked like a planet. It was later discovered that they were caused by a material ejected from a central star. This material is illuminated by the central star and shines, and an emission spectrum can be observed, the central star usually ends up as a white dwarf.

The nursery of planets

I was going to start talking about our own Solar System, but I see the need to talk about this very complex subject: the planetary genesis. How are they born? What is a planet? If they are born, do they die? What has already been considered as a common process, has been proving extremely chaotic. So small, so insignificant in the face of galactic clusters, the planets are very diverse. There is no celestial body that concentrates on itself, such a complex interaction between geological, chemical, and biological processes. Only on a planet can we shelter life as we know it.

Studying the planets is not easy at all, and it is not a single subdivision of astronomy that covers such a vast subject. For this study, scientists join astrophysics, planetary science, statistical mechanics, and nonlinear dynamics.

It all starts with an interstellar cloud that collapses.

We spoke earlier about the stars and about interstellar dust. All stars in the Universe have this disk of dust, which are tiny particles of ice, iron, rock, and other solid substances that condense into the cooler outer layers of stars and are ejected into the interstellar medium. This disk is the first and primordial condition for the formation of the planets. Thus, even the most powerful planets have small origins; are born from the ashes of a long-dead star and now another new one (another confirmation that we are star dust) the gases present are composed mainly of hydrogen and helium.

Organizing the mess

At this stage of creation, the small grains of dust will mix with the gas and begin to collide with each other. To ensure light reaches the darkest regions of the disc's interior, the grains intercept starlight and re-emit high-wavelength infrared light. The density, temperature and gas pressure decrease with the star's distance. Up to a balance between rotation, gravity and pressure and the gas rotates with a smaller speed around the star (when compared to an independent body at the same distance). At that time, the smaller grains are swept by the gas, because the larger grains exceed the velocity of the gas, which will try to decelerate it and cause it to spiral inward toward the star. The larger these grains become, the faster they spiral. As these grains approach the star, it warms them and water and other volatile substances evaporate. The distance at which this happens, called the ice line, is between 2 and 4 AU. The ice line separates the planetary system in an internal region, poor in volatile materials with rocky bodies; and an external region rich in volatiles, populated with icy bodies. In our Solar System, the ice line defines the boundary between the inner rocky planets and the outer gaseous giants. In the ice line itself, water molecules tend to accumulate as the grains evaporate. This production of water generates a discontinuity in the properties of the gas in the ice line, which leads to a pressure drop in that region. The balance of forces causes the gas to accelerate its rotation around the central star.

Consequently, the grains of the surroundings do not feel a contrary wind, but in favor, which increases their speed and prevents them from migrating inland. As the grains continue to reach the outer parts of the disc, they accumulate in the ice line.

Forced to join, the grains collide and grow. Some of them compensate by crossing the ice line and continue to migrate inland, but in the process, they are covered by semi retained snow and complex molecules, which make them bulkier.

Through these processes, the dust grains are compacted forming bodies with a few kilometers in extension, known as planetesimals. Almost at the end of the stage of planet formation, planetesimals have trapped virtually all of the original dust. A little more than a decade ago, scientists dedicated to the study of planet formation had to base their theories on a single example: our solar system. Today they have hundreds of mature systems and many others on the move. No two are alike.

Climbing the steps

Better than the expression "blind in shooting" is "astronomer in the early planetary". Collisions between planetesimals will serve to join or separate them. The balance between aggregation and fragmentation leads to a dimension distribution in which small bodies account for most of the area of emerging systems and large one's account for most of their mass. The orbits around the star may be elliptical over time, trailing gas and collisions tend to make it circular.

In the beginning, the growth of a body is self-sustaining. The greater the planetesimal, the greater the force of gravity it has and thus it trails its companions of lower mass faster. When these bodies reach a mass equivalent to lunar mass, however, these bodies exert gravitational forces so intense that they disturb the solid material of the neighborhood, distancing it before they collide with it. It is in this way that they control their own growth, that is how a "planetary oligarchy" emerges: a population of planetary embryos with similar masses that vie with each other for what remains of the remaining planetesimal embryos.

There is a centralized range in the orbit of each embryo, we call this range of feeding zone. When the planetary embryo aggregates most of the planetesimals that inhabit this range, the cycle of its growth ends. The size and direction of the feed zone tends to increase with the distance from the star. This oligarchic growth fills the system with an excess of potential planets, but only a minority of them will survive.

The birth of gaseous

It is probable that the planet Jupiter originated in a very small embryo, probably the size of the Earth, only later it accumulated an amount of about 300 terrestrial masses in gas. The prototype of the embryo feeds on gas from the disk, but when it is absorbed, the gas releases energy and when it stabilizes, it cools. As a consequence, the growth rate of the embryo will be affected by the time of the gas cooling. Thus, if this cooling is slower, the star can expel the gas from the disk even before the embryo has a chance to develop a dense atmosphere. The main bridge heat transfer bridge is in the flux of radiation through the outer layers of the emerging atmosphere, which is defined by the opacity of the gas (determined mainly by its composition) and by the temperature gradient (largely defined by the initial mass of the embryo).

Early models indicated that the embryos needed a critical mass ten times the Earth mass, so that it could allow a sufficiently rapid heat transfer. Large embryos like these may emerge near the ice line, where there would already be an accumulation of matter. Perhaps this is why Jupiter, the largest planet in the Solar System, is there. Embryos can also emerge anywhere else if the disk contains more raw material that planetary system experts often assume. In fact, astronomers have already observed many stars whose discs are sometimes denser than the traditional estimate, and in that case, heat transfer is not an insurmountable problem.

Another factor that acts against gaseous giants is that the embryo tends to spiral toward the star. In a process known as Type I Migration, the embryo shoots a wave into the gaseous disk which, in response, pulls the embryo's orbit inward gravitationally. The wave configuration follows the planet as the wake of a boat. The gas on the side of the embryo that is furthest from the star slowly turns the embryo itself trying to make it back, slowing it down, while the gas inside the orbit rotates faster, pushing the embryo out, accelerating it. The outer region, being larger, overcomes the tug of war and causes the embryo to lose energy and to approach the star of several astronomical units in about 1 million years. This migration tends to be blocked near the ice line, where the gas that blows like a contrary wind starts to blow in favor, giving additional impulse to the orbit of the embryo. The growth of the embryo, its migration and gas decrease are steps that occur at practically the same rate. Victory depends on chance. In fact, several generations of embryos can start the process, just to migrate before completing it. In its wake, heaps of new planetesimals from the outer regions of the disk head inland and repeat this mechanism until finally, its process is successful, a gaseous giant form, otherwise the gas is lost and no gaseous giant originates.

The balance between the processes depends on the characteristics of the original matter of the system. A third of the stars rich in heavy elements are orbited by planets with the mass of Jupiter. These stars probably have denser discs that gave rise to larger embryos and were able to escape from the heat transfer bridge. Conversely, few planets form around smaller stars or with fewer heavy elements.

Once started, growth is accelerated to a surprisingly fast pace. A planet with the mass of Jupiter can add, in about a thousand years, about 150 land masses. During this process, it dissipates so much heat that it can, for a brief interval of time overshadow the brightness of the star it surrounds. The planet attains its stability when it allows Type I Migration. Instead of the dust disk displacing the planet's orbit, the planet shifts the orbit of the gas from the disk. The gas inside the orbit rotates faster than the planet, causing the planet's gravity to try to bring the gas back, pulling the gas toward the star (and away from the planet). The gas outside the planet's orbit rotates more slowly, so it tends to accelerate it, shifting it out - again moving away from the planet. In this way, the planet creates an empty space, a gap in the disk and interrupts its supply of raw material. Gas tries to reoccupy this gap, but depending on the planetary mass, it cannot reoccupy this place.

A precise moment defines the critical mass: the sooner the planet forms, the greater its final size, because much of the gas remains in the feed range. The fact that Saturn is smaller than Jupiter, is that it simply formed a few million years after it.

The cosmic dance of the chairs

There is a curiosity that involves many extra-solar planets: many of them revolve around their stars in very tight orbits, much closer than the orbit of Mercury around the Sun. These planets known as "Hot Jupiter" could not have formed in their current positions, basically because the orbital feeding zones were too small to provide enough material. Its existence requires a third stage following events of the planetary genesis, which for some reason did not happen in our Solar System.

First of all, it is necessary for a gaseous giant to form on the outside of the planetary system, near the ice line, while the disc still contains a considerable amount of gas. This requires a strong concentration of solid material.

Then the gaseous giant has to move to its current position. Type I Migration does not handle this alone, because it acts on the embryos before they can add enough gas. In this case, Type I Migration does not occur, but Type II Migration occurs. The giant planet creates a gap in the disk and prevents the flow of gas through its orbit. In doing so, it must fight against the turbulent gas of the adjacent regions of the disk. The gas never stops flowing into the gap and its diffusion towards the central star forces the planet to lose orbital energy. This process is relatively slow, taking millions of years to displace a planet some astronomical units, reason why the planet must originate in the inner part of the Solar System to be able to finish orbiting the star. As he and other planets migrate to the interior, they drag in that path any planetesimals and emissions that have been left.

Third, we need a mechanism that deactivates migration before the planet is totally swallowed by the star. The magnetic field of the star must create an empty gas cavity in the vicinity of the star: without gas, the migration is paralyzed. There is an alternative that assumes that perhaps the planet can create tides in the star, and the star can cause a torsion in the planetary orbit. If this protection does not happen (and it is what happens in many systems) the planet is heading towards the bottom of the star.

Creating gaseous giants

If a gaseous giant can form, the formation of other planets is facilitated. In an analogy to our Solar System, Jupiter may have helped Saturn to form faster than he could have done on his own. Jupiter also had participation in the formation of Uranus and Neptune. Without it, these planets would never have reached their present dimensions in the distance from the Sun. Without any aid, the formation process would have been so slow that the disk would have dissipated before it was completed, leaving planets in the mid of training. Being able to count on an initial gaseous gas brings several advantages. At the outer edge of the gap opened by it, the material accumulates practically for the same reason that accumulates in the ice line, that is, a pressure gradient causes the gas to accelerate and act as a favored wind in the grains and planetesimals, paralyzing its migration from regions farther from the disc. Another important effect is that the gravity of the first gaseous giant tends to fling near planetesimals to the outer limits of the system, where they can form new planets.

The second generation of planets is formed from the material collected from the first gaseous giant. In the case of Uranus and Neptune, the accumulation of planetesimals was providential: the embryos became extra-large, about 10 to 20 terrestrial masses which delayed the beginning of accretion of gas - and from that point on, there is little gas to be aggregate. These bodies ended up accumulating only about 2 land masses of gas. They are not gaseous giants but cold giants and have been shown to be the most common type of giant planets.

Second-generation gravitational fields of planets represent an additional complication to the system. If the bodies form too close to each other, the interactions between them and the gaseous disk can capture them for new, highly elliptical orbits. In the case of our Solar System, all planets have nearly circular orbits and are sufficiently separated to guarantee certain immunities to each other's influences. In other planetary systems, however, it is normal for orbits to be elliptical. In some, the orbits are resonant (this means that some orbital periods are related by a ratio of small integers). It is quite unlikely that birth occurs under these conditions, but there can be a natural occurrence when the planets finally migrate, by mutual gravitational forces. The difference between these systems to our solar system may simply be in the initial portion of gas.

Many stars form clusters and more than half of them have partners in binary systems. The planets can be shaped in different planes of the stellar orbit. In that case, the gravity of the companion realigns quickly and distorts the orbits of the planets, creating systems that are not planar, like our solar system but spherical, in the same way as a swarm of bees fly around the hive.

A Swarm of Lands

Planetary scientists believe that planets like ours are much more common than gaseous giants. While the gestation of a gaseous giant requires a delicate balance between competitive effects, the formation of rocky planets is much more robust.

The four rocky planets - Mercury, Venus, Earth and Mars - are mainly composed of materials with high boiling points, such as iron and rocks rich in silicates, evidence that these were formed within the ice line and did not migrate significantly. Within this range of distances, the planetary embryos in a gaseous disk could reach dimensions of up to 0.1 terrestrial mass, nothing much beyond Mercury. If there were more growth, some orbits might collide, which would lead to the collision and fusion of some planetary embryos. This is relatively easy to explain: after the gas evaporates, the embryos will gradually destabilize their orbits, and after a few million years they will be elliptical enough to intercept.

What is a little more complicated to explain is how the whole system self-stabilizes again and how the terrestrial planets came to rest in the nearly circular orbitals we know today? Part of the remaining gas may be the clue, but to begin with, if the gas were present it would have prevented the orbits from becoming unstable. One idea is that, after the practically formed planets, there is still a relatively large amount of planetesimals. In the next 100 million years, the planets turned some of these planetesimals and diverted the remainder toward the Sun. The planets transferred their random motion to the condemned planetesimals and entered almost circular orbits.

There is another idea that the long range of Jupiter's gravitational force has caused the migration of terrestrial planets in formation, bringing them into contact with the new material. This effect would be more intense in special resonant sites which, over time, moved inward as the orbit of Jupiter assumed its present configuration. Radiometric dating indicates that asteroids were the first to form (4 million years after the formation of the Sun), followed by the formation of Mars (10 million years later) and Earth (50 million years later). If uncontrolled, the influence of Jupiter would have pushed all the terrestrial planets into Mercury's orbit. How was this outcome prevented? Perhaps the planets had become too massive for Jupiter to move them significantly, or they would have been driven out of Jupiter's range of influence by powerful impacts.

Despite this, many planetary scientists do not believe that Jupiter was decisive in the formation of rocky planets. Many Sun-like stars have no planets like Jupiter, though they still contain debris filled with dust, indicating the presence of planetesimals and planetary embryos that could join together to form worlds like Earth. How many systems display Earths and do not display Jupiterers?

The Earth passed a decisive moment between 31 million and 100 million years after the formation of the Sun, when a Mars-sized embryo collided with proto-Earth and launched into space enormous quantities of matter that solidified, according to a theory, it was this matter of the Earth itself that formed the Moon. Another effect of these gigantic impacts was the ejection of the tenuous primitive atmosphere. The present atmosphere of our planet originated mainly of the gas imprisoned by the planetesimals that clashed with him and later with the gases that the volcanoes were released.

Gathering the trash

By now the planetary system was almost complete. Some adjustments still occurred to complete the final adjustments: the disintegration of the larger cluster of stars, which could destabilize the orbits of the planets gravitationally; other instabilities developed after the star cleaned its last gas disk; and the continuous scattering of planetesimals left over by the giant planets. In the Solar System, Uranus and Neptune cast the planetesimals outward toward the Kuiper Belt, or inward toward the Sun. Jupiter, with its mighty gravity, pushed them farther toward the Oort Cloud, located in the confines of the gravitational domain of the Sun. The Oort cloud could contain material equivalent to 100 terrestrial masses. Sporadically, a planetesimal of the Kuiper Belt or the Oort Cloud moves into the Solar System in the form of a comet.

As the planetesimals spread, the planets themselves migrated a bit. This could explain the synchronism between the orbits of Neptune and Pluto. Saturn, for example, may one day have orbited closer to Jupiter and then shifted outward in a process that could explain late heavy bombardment - a practically intense period of impacts on the Moon (and probably Earth) that occurred around 800 million of years after the sun had formed.

The Solar System

The Solar System sits in one of the arms of the Milky Way. It is formed by the Sun the only star, and more than 1,700 smaller celestial bodies, among comets, asteroids, and planets with their satellites. The Solar System is constituted by a central star, in the case our Sun and the bodies that it orbits.

This system consists of 8 planets and their moons, asteroids and comets. The Solar System also contains interplanetary gas and dust. The planets that orbit our system are classified into two categories:

- Rocky planets: Mercury, Venus, Earth and Mars.
- Gaseous giants: Jupiter, Saturn, Uranus and Neptune.

Between these two categories of planets, such as a watershed, there is a strip containing thousands of pieces of rocks that orbit the Sun, which is known as the Asteroid Belt. Most planets orbit the Sun in the counter-clockwise direction (which is the counterclockwise movement when viewed from above).

The planets and the moons spin on themselves, while simultaneously revolving around the Sun, comets and asteroids also revolve around the Sun; while the entire Solar System revolves around the galactic center.

The sun

The Sun is the central star of the planetary system we inhabit. It is a star that is in its half-life phase, about 5 billion years old, and so it will continue to shine like today for another 5 billion years. The Sun is a yellow star of main sequence, with about 1.4 million kilometers in diameter. Its composition is almost entirely made of hydrogen and helium. It is in the core of the Sun that hydrogen is converted into helium by nuclear fusion, a process that releases energy. Energy travels from the nucleus through radioactive and convective zones to the photosphere (which is the visible surface), where it leaves the Sun in the form of light and heat.

It is in the photosphere that there are areas usually dark and relatively cold, that are denominated by sunspots, that usually appear in pairs or in groups, that are caused by the magnetic fields. Another type of solar activity is the eruptions; commonly associated with sunspots and lumps.

Solar flares are sudden discharges of high energy radiation and atomic particles. Already the protrusions are huge arcs or filaments of gas that extend through the solar atmosphere, some can last hours, other months. The solar wind originates from the light particles escaping from the crown (upper atmosphere) that travel through space at hundreds of miles per second. The chromosphere and crown are visible from the Earth when the Sun is totally eclipsed by the Moon. The Sun has an average Earth distance of 149,9 million kilometers. It has a rotational period of 25.38 days. The star surface gravity can reach 27.94 G and temperatures range from 1,577 ° K to 5,778 ° K.

Mercury

The closest planet to the Sun is Mercury, and it orbits at a distance of 53 million kilometers. Because of its proximity to the Sun, this planet moves faster than the other planets, traveling at an average speed of 48 km per second and completing its orbit in 87.97 days. Mercury is the smallest planet in the Solar System and is the first rocky planet. Most of its surface has been heavily bombarded by meteorites, although there are some regions that have not suffered so many impacts. The largest crater was named the Basin of Heat, which has a diameter of 1500 kilometers approximately. It is believed to have formed when an asteroid collided with the Mercurian surface and was surrounded by concentric rings of mountains caused by impact. The rupes are the ridges that the surface presents, perhaps these were formed when the hot core of the planet was still cooled and contracted, an event that occurred about four billion years ago, deforming the surface of the planet. The surface of Mercury has an exclusive feature, which is the cliffs and systems of crystals with a few kilometers of height and that extend for hundreds of kilometers. Mercury revolves around its axis very slowly, taking about 58.65 Earth days to complete its rotation. As a consequence, the solar day (interval between one sunrise and the next) in Mercury takes about 176 Earth days, twice as long as the 88-day year of the Mercurian year. Surface temperatures in mercury have extremes ranging from 700.0 K on the day side to 100.0 on the night side. At nightfall the temperature drops very quickly because there is almost no atmosphere on the first planet away from the Sun. This layer of gases comprises only a small amount of helium and hydrogen captured from the solar wind, but there are still some small traces of other gases.

The name of this planet derives from Roman mythology, which corresponds to the God of industry and commerce.

The observation of the planet, by virtue of its proximity to the Sun and its proximity to the horizon, takes place an hour or two before sunrise or after sunset. The main attraction of the telescope is the phases: in the lower conjunction of the Sun (which corresponds to the new Moon), is invisible, in the upper conjunction (phase that corresponds to full Moon) the disk is all illuminated. The best times of observation are in the proximity of the maximum elongations.

Venus

The second planet away from the Sun is Venus and is also a rocky planet. It has the longest rotation period of the Solar System, which is 243.3 Earth days. This is due to its inverted rotational movement: The planet rotates slowly around its axis and in a retrograde sense (imitating the movement of the hands of the clock when seen from above). Venus is slightly smaller than Earth, its diameter is 12103.6 kilometers and its structure is similar to our planet: a semi-solid metallic core, surrounded by a crust and a rocky mantle. After the sun and the moon, Venus is the brightest object in the sky and this does not happen by chance: its atmosphere intensely reflects sunlight, a dense atmosphere composed almost completely (about 96%) of carbon dioxide, an atmosphere that retains heat in a powerful greenhouse effect. Consequently, Venus is the hottest planet in the Solar System, with a surface temperature of 735.0 ° K. Sulfuric acid clouds travel around the planet surpassing speeds of 300 km / h. Although the planet takes 283.2 days to complete a rotation, high velocity winds cause clouds to travel the planet in just four Earth days. High temperature, acidic clouds and the enormous atmospheric pressure are compositions that make the environment extremely hostile. There are over 100,000 small volcanoes on the surface, alongside hundreds of large volcanoes. Volcanic flows have produced long winding channels that extend for hundreds of kilometers, one of which reaches 7 thousand kilometers in length. Images of mountainous regions over 2.5 kilometers, made by NASA's Magellan mission (1990-1994), are usually bright, which is characteristic of damp soil. However, there is no liquid water on the surface to explain this brightness. One theory suggests that the shiny material may be composed of metal complexes.

The name of the planet Venus comes from Roman mythology which means the goddess Venus, which was originally linked to vegetation and crops, later the goddess became known as the goddess of sexual love and beauty.

Its beauty is similar to that of Mercury, but because it is more distant than it can be observed for 2h30m or 3h, but always before sunrise or sunset. It is interesting to note that two "stars" popularly known as "Estrela d ' Alva "and" Evening Star "are, in fact, the planet Venus.

The Planet Earth

The third planet at a distance from the Sun is the planet we inhabit, our home, our home. And this is the largest and densest rocky planet in our planetary system. Our planet is the only rare jewel in the entire Universe that contains confirmed intelligent life. The rocky interior is metallic of Earth a trademark of the rocky planets; however, the earth's crust is peculiar, it is divided into plates which we named by tectonic plates, which are separated and move slowly between each other. In the limits of these plates, because of the clashes between them, volcanic activities and earthquakes happen. The terrestrial atmosphere acts as a protective mantle, blocking the harmful radiation rays and prevents the meteorites from reaching the surface of the planet, our atmosphere still accumulates heat and avoids extreme colds. About 70 percent of the Earth's surface is covered by water, which has not yet been found in liquid form on other planets. The Earth has a natural satellite, the Moon, which is sufficiently large that both are considered as a double planetary system. The Earth lies at a distance of 149.9 million kilometers from the Sun and has a diameter of 12,713 kilometers. Our planet takes 365.26 days to complete a revolution around the Sun.26 days to complete a round around the Sun.26 days to complete a round around the Sun.

In some of my lectures, I leave the last hour for my spectators to ask me questions that intrigue their minds. Some are easy to answer, others not so much, which I used to model this book. In one of these lectures, I remember that they asked me:

- What is the exact size of planet Earth?

- $(ab) / a = -$ I replied.

Everyone who was present was perplexed, so I had to answer in words what I had said in numbers:

"Our planet has an approximately spherical shape, but a little flattened at the poles. About the equation I used to answer, I'll explain. The equatorial radius, in the case represented by a, is 6,378 km. The polar radius, which in our equation is represented by b, has the measure of 6.357. The difference between the equatorial and polar radius corresponds only to that of the equatorial radius, which is due to the greater centrifugal force in the equatorial regions than in the polar regions.

Our natural satellite: The Moon

The Moon is the first natural satellite in distance from the Sun, and the only one that the Earth has. Its size is relatively large for a satellite, with a diameter of 3,476.3 km, which means almost a quarter of the diameter of our planet. The amount of surface we can see, depends directly on the fraction of the Moon is under sunlight, causing the phases of the Moon. Our natural satellite is dry and sterile, with no atmosphere or even water. It consists mainly of solid rock, although its core possesses molten rock and iron. The surface is dusty, with plateaus covered with craters caused by the impact of meteorites and plains in which large craters were occupied by solidified lava, forming dark areas, called seas. The seas occur mainly on the visible side, which has a thinner crust than the hidden side. This issue of the visible and invisible side almost always appears in the lectures I make on the subject.

Some people, do not know that we can only see one side of the moon, or know, do not understand why this occurs. I discovered this when I was 9 and today, as I write this book, I had the sad news that the person who taught me this fact - and here my sincere thanks goes to the Kemmsies family - passed away. I remember the walks we did together and he explained to me: The Moon takes 27.32 days to take a turn around itself and exactly the same time for a round around our planet; as a consequence of this fact, we always have the same face facing us.

Mars

The fourth planet away from the Sun is also called the Red Planet and is the outermost rock planet. In the 19th century, astronomers observed what they believed to be signs of life on Mars. These signs included marks with appearances of channels on their surface and dark spots that appeared to be vegetations. It is now known that the "canals" were optical illusions and the dark spots were areas where the red dust covering most of the planet was carried away by the winds. Fine dust particles are often charged with dust storm winds that occasionally obscure almost the entire surface. The residual dust in the atmosphere causes a reddish hue in the Martian sky. The northern hemisphere of Mars presents many extensive plains formed of volcanic lava solidifies, while the southern hemisphere has many craters and large impact basins. There are also several very large extinct volcanoes, such as Mount Olympus, which is 600 km in diameter and 25 km in elevation, the largest volcano in the Solar System. The surface also has many canyons and channels with branches. The canyons were formed by displacements of the surface crust, but the channels appear to have been formed by running water which has now dried. The

Martian atmosphere is much rarer than that of Earth. Mars has two tiny moons of irregular shapes, called Phobos and Deimos. Its very small dimensions indicate that they would have been asteroids captured by the gravity of Mars. The southern hemisphere has many craters and large impact basins. There are also several very large extinct volcanoes, such as Mount Olympus, which is 600 km in diameter and 25 km in elevation, the largest volcano in the Solar System. The surface also has many canyons and channels with branches. The canyons were formed by displacements of the surface crust, but the channels appear to have been formed by running water which has now dried. The Martian atmosphere is much rarer than that of Earth.

The name of the planet comes from the Roman mythology that means god of war, by its reddish color.

Mars was topographically mapped with great precision, a satellite orbiting around the planet Mars, the MGS - Mars Global Surveyor, uses a device called a laser altimeter from Mars' global surveyor. An MGS camera takes photos and images can be seen on the website of the company that built the camera: www.mssg.com .

The European Space Agency does not have as much publicity as NASA, but the Europeans have a satellite called Marx Express, which began to orbit the red planet on Christmas 2003 and fascinating images can be seen at: www.esa.int/Spcials/mars_Express .

Jupiter

Jupiter is the fifth planet away from the south and the first of the four gaseous giants. Of all the planets in the Solar System, this is the largest and most massive, with a diameter of 142,984. Jupiter is believed to have a small rocky core encased in an inner mantle of metallic hydrogen. The rapid velocity of rotation of Jupiter (0.41 days) makes the clouds of the atmosphere form belts and zones that circulate the planet parallel to the equator. The belts are layers of dark, low, relatively warm clouds; the zones are layers of bright, high, and cold clouds. Between the belts and the zones, there is turbulence that causes structures of white oval clouds and red spots, in both cases, huge storm systems.

The most notable structure is a storm called the Great Red Spot, which comprises a spiraling column of clouds three times the Earth, which rises about five miles above the upper layer of clouds. Jupiter describes an orbit with a small eccentricity, and spends 11 years and 86 days to go around the Sun. Jupiter has a thin main ring, inside which there is a thin halo of fine particles, which extends towards the planet, but they are dark and invisible to amateur telescopes.

By the time I write this book (second half of 2017, early 2018) Jupiter has 64 satellites. The first four satellites, major and largest, were discovered in 1610 by Galileo Galilei and are now known as Galilean moons: Calisto has a dark surface, marked by many white craters. The surface is probably dirty ice - a mixture of ice and rock. The impacts of asteroids, comets and large meteoroids exposed the clean layer of ice, thus forming the white craters. The surface can be a frozen layer that covers an ocean of water and mud, perhaps 150 km deep. Europe is the only place in the Solar System outside the Earth where scientists have strong evidence that there is liquid water. With 5,264 kilometers in diameter, Ganymede is the largest moon in the Solar System (surpassing the size of Mercury with 4,881 km). The surface of Ganymede consists of light and dark terrain, perhaps ice and rock, respectively. The most notable brand is Valhalla, a huge basin of impact in the size of the United States in ring form. The surface of Io is dotted with more than 80 active volcanoes. This moon is the only place beyond Earth where there is definite evidence of volcanism.

Recent discoveries reveal that there have been diamond rains in Jupiter's atmosphere. The name of Jupiter also comes from Roman mythology and alludes to the supreme God.

Saturn

This is the sixth planet at a distance from the Sun. It is a gaseous giant almost as large as Jupiter, with a diameter of about 120,536 km. It is believed that Saturn is composed of an inner mantle of metallic hydrogen. Around this mantle there is a mantle of liquid hydrogen immersing in the gaseous atmosphere. The clouds of Saturn form belts and zones similar to those of Jupiter, but more obscured by cloudiness. Storms and swirls that look like red and white spots occur in the clouds. Saturn has an extremely thin but broad ring system, which is less than one km thick, but extends for about 820,000 km. The main rings are composed of thousands of narrow rings, each made up of fragments of ice ranging from fine particles to large pieces several meters in diameter. Some rings are seen from Earth only using a binocular. Saturn has 200 known moons, strangely, some of them are co-orbital, that is, they share an orbit with another moon.

The observation of Saturn is not difficult, just as the planet Jupiter is easily visible to the naked eye in the sky, like a star of first greatness. Its rings were first observed by Galileo Galilei, but were only studied by C. Huygens in the seventeenth century.

Uranus

Uranus is the seventh planet of the Solar System and the third in size, with a diameter of 51,118 km. Discovered in

1781 by William Herschel, it was originally called Georg Um Sidos in honor of the English sovereign George III. A day in Uranus lasts about 17:14. Like Jupiter and Saturn, the rotation of this type of planet is much faster in analogy to the rotation movement of the Earth, in contrast, its movement of translation is much greater; in consequence of their distance. The seventh planet takes 84 Earth years to make its journey around the Sun. A feature that is Uranus' trademark is its equator: Unlike most planets, its equator forms almost a right angle to its orbital plane, that when we use as a point of reference the directions here of the Earth, the equator of Uranus begins at its north pole and ends at its south pole. Sometimes the north pole of Uranus points toward the direction of the sun and for our planet and once it is the south pole that points us toward our direction. Uranus also has rings, which are composed of dark materials, probably carbon-rich rocks, parts of meteorites known as chondrites and carbonaceous. The moons and rings of Uranus orbit him in relation to his equator.

So far 27 Uranus moons have been discovered orbiting its equator. The seventh planet has an impressive distance from the Sun: about 3.0 billion kilometers. All the moons of this giant, gaseous, icy planet are made of ice and most orbit out of the rings. The final moons inside the rings are small and dark, with diameters less than 160 kilometers, while the outer moons have diameters ranging from 470 to 1600 kilometers in diameter.

Neptune

Neptune is the farthest planet from the Sun, is at an average distance of 4.5 billion kilometers, with a diameter of 49,528 kilometers. It is believed that this planet is composed of a small rocky core, and it is surrounded by a mixture of liquids and gases. Its atmosphere has an immense diversity of characteristics that are remarkable in the clouds: the biggest one is the Great Dark Spot, which is as big as the Earth, the Little Dark Spot and the Skate. The Small and Large Dark Spot are huge storms moving around the planet with winds of 2,000 KM / h. Already the Skate is a large area in the Neptunian sky of formation of clouds of cirrus. Neptune has a set of rings and 13 known moons. The largest of these is Triton, which has 2706.8 kilometers in diameter, and this one has a particularity: it is in Triton where the lowest temperature of the Solar System is found, only 38.0 K. And even more, unlike all moons in our Solar System, Triton orbits its planet in the opposite direction to the rotation of Neptune. This moon formed of ice and rock had its surface formed by cryovolcanic, which means eruption and flows of substances very icy and, instead of hot, icy and melted.

And Pluto?

I learned in school that the smallest planet and also the farthest from the Solar System was Pluto, which had a diameter of 2,322 km and its moon, Charon, had a diameter of 1,207 km. But today it's very different. In the year 1999, a group of astronomers attempted to lower Pluto to the status of asteroids, others believed that that foreign body with an eccentric orbit should be only one of the major aspects of the Kuiper belt. But ordinary people and other astronomers have won, arguing that Pluto is round as a planet, has a large moon and has been considered a planet since 1930, when American astronomer Clyde Tombaugh discovered it. Pluto was considered a planet far too far away to be observed, however, with the recent advancement of observation equipment we discovered three more moons. We find that its composition is rocky and its climate is cold and the ice there is real and not melted, as it occurs in gaseous planets like Uranus and Neptune.

Another interesting thing happens when we talk about Pluto and Charon, its greatest moon: Pluto's rotational period is 6.36 Earth days and in Charon, this period is exactly the same time. Making a comparison with our satellite, the moon, a hemisphere of the moon is always visible to our planet, but it does not happen from the same hemisphere of the Earth facing the moon. In Pluto this is totally impossible and in Charon also: anyone on the planet only sees the same hemisphere forever and whoever looks at the moon will see only one half of Pluto.

Here's the climax of the show: in the year 2006 after the discovery of a KBO (Kuiper Bell Objects, greater than Pluto), called Eris. This discovery precipitated the debate in the International Astronomical Union (IUA) that in 2006 approved a resolution that defined planet as a "planet: a planet that orbits the Sun, has a spherical shape and is the body that dominates gravitationally their neighborhood. It was with this new definition that it eliminated with the demotion of the planet's status and today it is denominated like planet dwarf.

Asteroids, comets and meteoroids

Despite the time (it's three in the morning), I find myself very excited to write about it: I've always been fascinated by the history of comets. I spent a lot of my teenage years researching history, lecturing, lecturing on astronomy and archeoastronomy. When we talk about comets, what fascinates me is the haze that surrounds them and the fascination they carry. I remember a lecture with the name that I opened this topic, which I did on February 16, 2013, the day after the meteor that crashed in Russia. I know it's been a long time, but I still remember the first question they asked me, which was: "What's the difference between asteroids, comets, meteoroids, meteorites?" In reality, they are all spatial rocks.

Beginning with the asteroids, these are celestial bodies that have irregular shapes and are usually located between the orbits of Mars and Jupiter. These bodies are rocky and can measure about 1000 kilometers in diameter, although most are smaller. Comets originate from a huge cloud called the Oort Cloud, which lies within the confines of the Solar System. These bodies are made up of frozen gases and dust, counting only a few kilometers in diameters. Comets have very elongated orbits and when they are diverted and placed in a long elliptical trajectory around the Sun, it moves away and then approaches again. When this happens, its surface begins to vaporize under the action of heat, producing a shiny and shiny wig (an immense sphere of gas and dust around the nucleus), one tail of gas and one of dust. Meteoroids are small pieces of rock or rock and iron, some of which are nothing more than fragments of asteroids or comets, ranging from tiny particles of dust to objects tens of meters in diameter. When we talk about meteors and meteorites, and this is an issue that I hear a lot in my lectures, it's the difference between them. Well, it works in this order: if a meteoroid enters the atmosphere of the Earth, and because of the friction with the atmosphere it fragments, in this case are meteors. However, there are cases where these space objects are a little larger, and with a rarity greater hit the ground with an amazing speed, which can reach 40 thousand kilometers per hour, these are called meteorites. them. Well, it works in this order: if a meteoroid enters the atmosphere of the Earth, and because of the friction with the atmosphere it fragments, in this case are meteors. However, there are cases where these space objects are a little larger, and with a rarity greater hit the ground with an amazing speed, which can reach 40 thousand kilometers per hour, these are called meteorites.

Some meteorites may exceed 10 or more times the destructive force of the atomic bomb that was dropped on Hiroshima in 1945. These meteorites often fall to Earth, but luckily most of them fall into the oceans; which correspond to 70% of the earth's surface. One of the frequent questions I hear in my lectures is why meteorites fall more in the northern hemisphere than in the southern hemisphere. The answer is simple: the northern hemisphere has much more land than the southern hemisphere, even though a good part of it is uninhabited, most are there.

When I had presented about 60% of my lectures, I associated comets with ethnoastronomy, and everyone was wide-eyed and open-mouthed. I'm not the best of the lecturers on astronomy, archeoastronomy and ethnoastronomy, but I know how to do it very well. I watch my audience's level of interest from sneezing, commenting, and the rattle of chairs. And in order to arrest my readers, I have always used phrases of effect, and when I came to the part where it said: "when the fate of men is ruled by the heavens", ready! Everyone was in complete silence.

I can bet you, dear reader, have probably seen some Hollywood movie that has fun worshiping the destruction of the planet. I can even cite a few here: two classics from 1998 are Armageddon and Deep Impact. But this type of film is not only American: Melancholia, awarded in 2011, which was directed by Lars von Trier, who is Danish. Death comes from space, directed by Paolo Hensch and is Italian.

But surely the most famous story began in the year 1978, where Richard Donner first told the story of Superman in the cinema: the rock that was totally scary was actually the involvement of an alien orphan that would come to become the greatest hero on the planet.

It is interesting to begin this part by telling the origin of the word "disaster", which comes from the Latin "dis-astra", which means against the stars. Our history here on Earth changes according to what happens up there: empire fall, robberies, murders, suicides, decimations, wars ...

Until the Middle Ages, the passage of comets was interpreted by astrologers as a sign of bad omens. Until that time, they blamed for the droughts, falls of kings, invasions enemy, all because of the passages of these stars; but this was pure charlatanism. Only in the European Renaissance did astronomy become a study.

In the year 1066, on the eve of the Battle of Hastings, a battle in which William II, Duke of Normandy, attempted to invade England: Halley passed and British troops regarded him as a bad sign. Result of the battle: the British lost the confrontation and Harold II lost his head.

In 1222, the Mongol emperor Genghis Khan was so impressed to see Comet Halley that he considered him his personal star. Genghis Khan interpreted it as a good sign for the invasion of Europe, which resulted in the deaths of thousands of people.

In 301, once again Halley came to fascinate the human mind, now it was the turn of
Giotto, famous Italian painter: this one painted the Crèche of Jesus Christ surmounted by Halley himself. In fact, there is a certain possibility that this comet may be the famous star of Bethlehem, the time that Jesus was born was the time of his passing.

The last appearance of Halley, that in 1910, caused hysteria there in the United States. The information that the tail of the comet contained toxic substances that would poison the earth made people go crazy. It was not small the number of people who raised large sums selling masks, which guaranteed to seal such substances. There are reports that many people locked themselves indoors, sealing doors and windows and died asphyxiated. However, the most famous account - as in the good American police fiction books - happened in Oklahoma, where a religious group tried to sacrifice a virgin to placate the comet and avoid contamination of the atmosphere. Of course, this did not happen: the virgin was saved by two policemen who made the local round. There is another account, also American, that a small locality gathered all the food and drink and indulged in an orgy feast, and when the food and drink was gone, they all went back to their homes, Halley passed and did no harm to the planet.

Making several requests: meteor shower

As I mentioned before, meteors are space rocks that enter the atmosphere. Those flashes that we see at night and commonly call "falling stars" are due to the fragmentation of these bodies in action of the friction with the atmosphere.

When we are far from the big cities and their lights, in good weather conditions and on a night without a moon, we can see some meteors: about five or six, with a bit of luck until a little bit more. However, on certain specific dates of the year, we can see 30, 50, up to 80 meteors per hour! Yes, that's right, 80 meteors an hour or more, a real meteor shower; and that's exactly what we call those exact times: meteor showers. These rains happen whenever the Earth crosses with the orbit of some comet and the trail left by them enters our atmosphere.

Meteor showers are usually named after the constellation from which their radiant comes out, that is, a specific point the sky from which the rain seems to be coming out. For example, the most famous meteor shower is that of Perseids, which at its peak produces up to 80 meteors per hour: this rain receives this name because the meteors that compose it appear to leave the constellation of Perseus.

The first meteor shower of the year is called Quadrantids, and that's a very odd name. But when they named this rain, they relied on a nineteenth-century stellar map, and it was with its radiant located in a constellation that is no longer recognized by the International Astronomical Union. Today to find its rain just look at a few degrees' northeast of the constellation Boötes.

The last meteor shower of the year is known as Geminids, and instead of being associated with a comet, is associated with an asteroid. However, scientists suspect that this asteroid is nothing more than a former comet, which no longer emits a head and a tail.

Observing the meteor showers

Advice and tips are never too much and here I will give good tips. For those who want to enjoy watching and counting meteors, the tip is to recline on a lounger chair (or lie on a rug with a blanket and a pillow, but if you have not slept well you can sleep in the middle of the show). You will need a watch, a notebook, a pen or pencil, and a flashlight. The best light for the lanterns is red, which is weaker and does not interfere with the adaptation of your eyes to the dark. In case you do not have flashlights with reddish lamps, you can adapt your painting your lamps with red nail polish. Although it is what everyone does, you do not necessarily have to look at the radiance of the meteor shower to enjoy it. Meteors fall all over the sky, and their visible paths can begin and end away from their radiant. Take a thermos with coffee, hot chocolate or tea and good meteor shower. So, you do not miss any great meteor showers, here's a list of the hourly meteors and rates.

The biggest meteor showers		
Rain name	Approximate date	Meteor rate (P / h)
Quadrantids	3 - 4 of January	90
Lipids	April 21	15

Eta ACUARIIDS	4th - 5th of May	30
Delta Acuariids	28 July 29	25
Perseids	August 12th	80
Orionids	October 21st	20
Geminids	December 13th	

The End of the Universe

Today, cosmologists work to present the fascinating setting of the early universe. And he is there, we can observe him with small telescopes or with the giant devices, which are kept by observatories all over the world. If it is already difficult to talk about something that has already happened, then imagine speculating on something that is still going to happen. Will the day come when there will be an end or our Universe is eternal? Einstein believed that the Universe had to be immutable, even if his equations showed him the opposite. If we think about it, how can we accept that at some point everything can simply disappear? Well, everything is entirely connected to its mass: it is it that will determine whether the Universe will expand forever or, at some point, will collapse at the end. After all, all the questions of the Universe depend on the mass, from birth to the death of any cosmic object depends on the mass. If we could calculate the mass of the Universe, we would have that answer. The fact is that we lack a data that is very precious: so far, scientists do not know precisely what is the actual amount of matter that exists in the Universe and could stop the expansion. Scientists have calculated that to stop the expansion, we would need an amount of 3 atoms per cubic meter, which is what we might call the critical density of the Universe. In case the average concentration of matter is below this critical density, it should expand forever, but if it is higher than that value, at some point our Universe will contract. We would have that answer. The fact is that we lack a data that is very precious: so far, scientists do not know precisely what is the actual amount of matter that exists in the Universe and could stop the expansion. Scientists have calculated that to stop the expansion, we would need an amount of 3 atoms per cubic meter, which is what we might call the critical density of the Universe. In case the average concentration of matter is below this critical density, it should expand forever, but if it is higher than that value, at some point our Universe will contract. We would have that answer. The fact is that we lack a data that is very precious: so far, scientists do not know precisely what is the actual amount of matter that exists in the Universe and could stop the expansion. Scientists have calculated that to stop the expansion, we would need an amount of 3 atoms per cubic meter, which is what we might call the critical density of the Universe. In case the average concentration of matter is below this critical density, it should expand forever, but if it is higher than that value, at some point our Universe will contract. Would require an amount of 3 atoms per cubic meter, that is what we could call the critical density of the Universe. In case the average concentration of matter is below this critical density, it should expand forever, but if it is higher than that value, at some point our Universe will contract.would require an amount of 3 atoms per cubic meter, that is what we could call the critical density of the Universe. In case the average concentration of matter is below this critical density, it should expand forever, but if it is higher than that value, at some point our Universe will contract.

We have a very serious and invisible problem: dark matter. Recent studies show that about 90% of what actually exists in the Universe is, in our eyes, invisible and undetectable. This is one of the opportunities that makes scientists not know how much matter is in our Universe. This matter can be in the form of dead stars, primordial black holes and particles predicted by the Theory of Supersymmetry and that until today have not been observed.

However, the Universe does not have to contract to end. There are a myriad of theories that point to some distinct purposes for what we call the cosmic house.

The Cosmic Death of the Universe

In the year 1956, the German physicist Hermam Von Helmholtz concluded that our
Universe was dying. To support this hypothesis, Helmholtz used as a principle the Second Law of Thermodynamics. This law says that by assuming an isolated system, thus not receiving energy from other bodies, the heat always flows from a warmer body to one that is colder. In this case. There is a fundamental direction for the flow of heat; we can say that the heat flow is unidirectional. This flow is commonly represented as an "arrow of time," which passes from the past to the present and indicates that this process is irreversible. Then we come to a physical concept called entropy, which is a characteristic property of the irreversible changes that can occur in thermodynamics. The second law of thermodynamics says that the entropy of an isolated system never diminishes, for if it did, we would have heat flowing spontaneously from a colder body to a warmer body.

The entropic concept can be generalized to all closed systems and entropy never decreases. Let's use the concept of entropy in our Universe as we know it: can we consider it as a closed system?

Maybe so, because there is no point in talking about the outside of the Universe. I bet you've never heard of a star system that was located outside the Universe. It can be seen that the entropy of it never really decreases, and on the contrary, it always grows. We will exemplify with the sun: it emits its heat into cold space, and that form of energy never returns, a process that is absolutely irreversible.

Maybe not. Can entropy be extended forever? This is a point of interest. If all stars emit thermal energy into cold space, and as we know that this stellar energy is not inexhaustible, that means that at some point the whole Universe will be the same temperature. When this moment happens, it will have reached what physicists call "thermodynamic equilibrium," which is the ultimate condition.

When we speak of a Universe that expands forever its final destiny would be, forever, full of radiation. Everything in it would somehow decay back into the radiation and this could happen through various processes. Some Great Unification theories allow the so-called "proton decay," which would be enough to annihilate all matter in the Universe. It is necessary to remember that the proton is not a fundamental particle and yes, composed by three quarks. These quarks are all the time interacting inside the proton and it could be that at any given moment, they would get close enough that gravitational interaction between them would increase. In this case, they could come together and form a mini black hole. Therefore, the proton would collapse due to its own gravity by virtue of the quantum tunneling effect that would occur within it. This mini black hole is highly unstable and disappears almost instantly: this process would give rise to a positron, which is the antiparticle of the electron, and would also give rise to a neutral pion.

If protons are really unstable and decay, even if this occurs over a long period of time (about years, which is the least of them) the consequences are tragic for matter in the Universe. Everything in it would be unstable and would eventually disappear at some point, as long as the Universe has enough time to reach the final stage. All celestial bodies, all existing forms of matter, would lose their protons through this process at some point in their lives.

With the decay of neutrons (neutrons are also formed by quarks) and protons, the universe would consist of positrons and pions. The pylon for being unstable would decay into two photons or an electron-positron pair. See, therefore, that the Universe has acquired more and more positrons; once all the protons have decayed into positrons, there will be an almost equal mix of them and electrons in the Universe. Initially, the electron and the positron will combine in a kind of mini atom, which is called by positronium, united by a mutual electric attraction, since they have different electric charges. This system is unstable and its orbital movement has the form of a spiral. This causes these particles to end up annihilating themselves. The time required for this annihilation depends on the initial distance between the electron and the positron. Estimates show that it would take years for positronia to form, and its constituents would have orbits of the order of many millions of light years. Inevitably, all postironies will, at a given moment, be annihilated.

Another process that would happen simultaneously would be the formation and subsequent evaporation of black holes by quantum processes. Stars such as white dwarfs and neutron stars could undergo processes that would turn them into black holes with their consequent evaporation. Note that the processes that are happening in a universe that expands forever, happen more and more slowly. Our Universe, in eternal expansion will have as final matter the cosmic background that has always been present, the radiation of protons and neutrinos that was created with it. The ordinary matter of the Universe will have disappeared and all black holes will have evaporated. In this distant future the Universe would be just an incredibly diluted "soup" of protons, neutrins and in number less and less of electrons and positrons that move away more and more slowly.

It is not yet known why the Universe could interrupt this thermal degradation and return to a process of matter creation. If this hypothesis is correct, it will take many, many centuries to happen, but it will be inexorable if the Universe expands forever.

Back to square one

We present in the previous topic a hypothesis for the final cosmic destiny based on the eternal expansion of our Universe. However, what if something happens and the expansion is stopped? What if there was a collapse? Obviously it would take a while for us to realize what would be happening, since this re-collection would occur very briefly.

A number of factors could show us that the Universe began a return, a first factor even if it was subtle. The first that would be a likely candidate for this factor would be the increase in cosmic background radiation (now 3K). This radiation is reminiscent of the Big Bang, the great initial explosion of the Universe, and this radiation cools as it expands. If by chance this temperature rise is detected, this would be a very important factor on the universal collapse.

Another factor that could also show this collapse would be the Redshift of galaxies. These shifts to the red of the spectral lines of the galaxies would begin to be replaced; rather than red, would be shifted to the blue, Blueshift. Clusters and clusters of galaxies begin to contract gravitationally, galaxies, in view of this concentration, begin to gather more and more within these clusters and clusters of galaxies. This will lead to increasing gravitational interaction, which would increase the number of collisions and disruptions of these galaxies within these clusters. Several gravitational bonds begin to release energy as the clusters of galaxies, their galaxies and their stars consequently join together progressively, and this accelerates the process of concentration of the Cosmos. However, this implacable tendency to contraction can be delayed by other physical factors existing in the universe, among them we can mention: rotation, nuclear energy and the enormous scale of astronomical systems, which causes these processes to happen slowly, which delays what is inevitable , which is the victory of gravitational interaction.

The gradual increase in gravitational interactions, the central regions of the galaxies - which are usually formed by huge black holes - will become immense and impatient star-devourers, which each mass swallowed, more mass adheres to itself, causing several "holes "In space-time.

According to this hypothesis, we are making a journey to the beginning of our Universe. As it collects, the temperature of the cosmic background radiation increases inevitably. All the celestial bodies, the Sun, the Moon, the stars and even our Earth will suffer the consequences. This radiation would become so intense that the night sky would turn reddish, and our planet, like all others, would not withstand the heat and be destroyed. The space would be filled with increasingly hot gases. The temperature would rise to over a billion degrees.

With the thermometers high up, the atomic nuclei would disintegrate, and be crushed, and the Universe would consist only of black holes. Protons and individual neutrons also cease to exist, and what would remain would be just a "soup" of quarks, its basic constituents.

The collapse will continue to accelerate. In this stage, black holes merge, gravitation dominates everything, which causes the curvature of space-time to become more and more accentuated, which causes it to become more and more compressed. Point, here lies our Universe.

Even more so, the collapse continues, more and more intense, until it leads to a great crushing, an implosion that swallows everything that remains. The final stage would be a huge fireball, similar to the initial singularity, that fireball that started our Universe 13 billion years ago.

The end may also be so. The end can be in many ways, just as it was the beginning, so the end still seems to us an unknown one that we need to answer. What would happen next? Of course we can speculate on many things. Who can answer the questions that will happen after what is likely to happen? Maybe a new explosion will happen and give rise to another Universe, what assures us that this has not happened before? Would that be possible or not? It's comical to finish an answer book with questions, but it's them that move the world. First of all, it would be necessary to discover and understand the processes that originated in our own Universe, processes that seem to be guarded by a greater force. If another Universe were to originate, would it be the same as ours? We can not answer yet. But science is striding forward, and before humanity reaches the middle of our century, we will have all the answers to these questions, and others that persist in populating the imaginary of men.

Conclusion

The starry sky always exerted fascination and perplexity to men, and this looks at the heavens since it is understood by people; thus crowning astronomy as the oldest of the sciences. It is not small the number of poets, mathematicians, philosophers, writers, thinkers, founders, a multitude of people were influenced by the perfection and rapture caused by the night sky. Night observation was one of the pillars of civilization's origin, the cornerstone: it was necessary to know the cycle of the seasons for the development of agriculture, and without agriculture, civilization would hardly be conceived. For social development, the people need minimum guarantees for their continued existence in one place. There is moreover, the need to measure time with a certain precision, and then astronomy enters again to help, it gave humans as a gift beyond the measurement of time with the Sun (days), the Moon (months) and the stars (years), still gave the possibility of geographic fixation. After mankind adopts this custom of observing the celestial objects, it begins to gradually abandon the nomadic life and begin the buildings of cities. On the other hand, astronomy also played a fundamental role in the search for what was unknown: without astronomy great navigations would not be possible and without them the discovery of new lands would not have happened either

Looking at the sky creates in our imagination a portal, this portal that leads us to other societies, other beings, other lands, think of others and forget about ourselves.

After much looking into the sky our vision has been changing, and we have created the ability to look and see, see and understand. We believed that the Earth was the center of the Universe, then we tried to place the Sun at the center of the Universe. Until the Italian appeared Giordano Bruno with his idea that the Universe could not have a center. How many people died defending new ideas? How many people suffered persecutions, only because they understood before others what was before their eyes? Today every month a new discovery comes up and it just figures in the Science or National Geographic magazines and it's all right.

We still try to understand what there is after the black hole event horizon. We still want to know what happened so that our own Universe would arise. We want and need to know where we are going.

Many discoveries have marked and changed our vision. When in the year 1929 Edwin Hubble discovered that the galaxies were moving away from us. Until this year the general belief was that the Universe was stationary and it fell. Before, they believed that the stars did not evolve, nobody dared to imagine that those bright spots of the night sky also died. We know today that they explode, and it is because of their explosions that other stars and other planets are formed. Life results from death, so that death is the result of life. Interesting and intriguing is that even before Hubble proved his theory and created his universal constant, Einstein had already understood and discovered it. In his calculations on the theory of general relativity he understood that this was a clash of reality, overturning a belief that has stood for millennia of years: when Hubble postulated his idea, Einstein understood his mistake: try to adapt his equations to fit his ideas. An evil that has become a trademark in history.

In the year 1905 Einstein made a precious discovery: space and time formed the same thing, space-time. One more belief came down. In this case, the time to be absolute: the clocks may even be the same, but the motions are different, and everything depends on the speed at which it moves: time becomes relative, it passes faster or slower. This was an interesting discovery for us to understand the Big Bang. This theory got a better understanding because we understood that we could move in time, that it is not static and not even immobile.

We know that at a given moment the Universe began to inflate, but before this everything was gathered in a single point, that singularity or cosmic egg that gave birth to all of us and everything that surrounds us. An old thinker said: There was a day when there was not yesterday. It is true. There was the day of the genesis, where all that had to be created, it could have been seven days, a thousandth of a second, who can say? After all, time is relative. When we speak of this stage in the age of the Universe, we can only speculate, since understanding belongs only to metaphysics and theories.

The future is the expanding Universe. The galaxies moving away, the stars moving away. This will generate future problems: because of the enormous distances and distances that will affect our Universe, we will arrive at a certain point where a new generation of stars can no longer be born, the force of gravity will not be so strong anymore. Although there are a myriad of theories that explain the end, in one part they combine: the scenery will be cold and dark, all matter will be swallowed by huge black holes, and even they will slowly evaporate. It's a bleak end. It is the phosphorescence that appears in the sea, which illuminates everything, and gradually fades away, until darkness takes over everything again, as it was in the beginning and will also be in the end, the Universe drags itself until the day there will be no tomorrow.

Glossary

Absolute time: Newtonian conception of time as universal with the consensual notion of the simultaneity of events, as well as the universally accepted time interval between two events; refuted by Einstein.

Acceleration: This is the reason an object changes.

Accuracy disk: a disc of matter that spirals around an object due to gravity.

Antimaterial: Matter composed of particles with mass and spin identical to those of the particles of common matter, but with opposite charge.

Antiparticle: Each species of particle of matter has a corresponding antiparticle. When a particle collides with its antiparticle, they annihilate, leaving only energy.

Aphelion: The farthest point a body reaches in its orbit around the Sun.

Apogee: The farthest point of Earth in the orbit of the Moon or an artificial satellite around our planet.
Asteroid: Small body that orbits the Sun, usually is located in the range of asteroids, between Mars and Jupiter.

Astronomical unit (UA): Unit of distance equal to the average distance between the Earth and the Sun: 149,597,870.

Atmosphere: The outer gas layer around a planet, satellite or star and there is no properly defined boundary, only becomes more and more rarefied until arriving at the space.

Atom: The basic unit of ordinary matter, consisting of small nuclei (formed of protons and neutrons) surrounded by electrons in orbit.

Axis of rotation: Imaginary line around which the body rotates. The axial slope is the angle between the axis of rotation and the perpendicular to the orbital plane

Big Bang: The great explosion that gave rise to the Universe in which we live.

Big Crunch: Singularity related to the end of the Universe.
Binary Star: A pair of stars where one orbits the other. About half of all known stars belong to groups of two or more

Black hole: A region of space around a collapsed star, where gravity is so intense that nothing, not even light can escape.

Bosons: Elementary particles, among which are the photons, gluons, intermediate vector bosons, and gravitons, which carry the four forces in nature.

Brightness: The brightness of a luminous body and is defined by the total energy radiating at a given moment.

Celestial sphere: An imaginary sphere on which celestial objects seem to be glued when viewed from Earth. The celestial equator is marked by the projection of the terrestrial equator on the celestial sphere. The celestial poles are points under the celestial sphere aligned to the north and south poles of the Earth.

Comet: Small body that orbits the Sun and in eccentric orbits.
Cone of Light: A surface in space-time that signals the possible directions for the rays of light passing through a given event.

Coordinates: Numbers that specify the position of a point in space and time.

Cosmology: The study of the Universe as a whole.

Cosmological constant: Mathematical resource used by Einstein to give the space an intrinsic tendency of expansion.

Day: Time interval that a body takes to complete a rotation around its axis. A sidereal day is the time it takes for a star to return to the same position in the sky. A solar day is the time interval between one sunrise and the next.

Doppler effect: Change in wavelength - whether sound or light - emitted by a moving body, noticeable when the source of sound or light is approaching or moving away from an observer. (If the source of the waves is approaching the observer, the frequency of the wave increases and the wavelength is shorter, producing sharp sounds and bluish light - the so-called deviation to the blue. , the wave frequency decreases and the wavelength is longer, producing bass sounds and reddish light - the so-called redshift.).

Eclipse: Total or partial obscuration of one celestial body by another. In the solar eclipse, the moon passes between the sun and the earth, hiding it partially or totally, but only visible in a small part of the Earth. In the Lunar eclipse, the Earth is between the Moon and the Sun.

Ecliptic: Plane in which the Earth orbits around the Sun.

Effect of time dilation: Delay of a moving clock as seen by a stationary observer; postulated by Einstein in his special theory of relativity. (At the slow relativity speeds of today's normal travels, this effect is negligible, at speeds close to that of light, however, time becomes appreciatively "slower." At the speed of light, time would be paralyzed.)

Electric load: The property of a particle by which it repels (or attracts) other particles that have similar (or opposite) signal charge.

Energy conservation: Law of science that postulates that energy (or its equivalent in mass) can not be created or destroyed, Famous law enunciated by Antoine de Lavoisier.

Entropy: A measure of the degree of disorder, or tendency to collapse, in any system.

Field: Something that exists in space and time, as opposed to a particle that exists only at a given time

Galaxy: A set of stars, gas, planets, and other bodies that are connected by gravitational force. The galaxies are classified according to their visual morphologies, they can be spiral, elliptical and irregular. They usually occur in groups known as agglomerates.

Giant and supergiant stars: Large stars with high luminosity. The giant stars are 10 to 100 times larger than the Sun, while the supergiant are the largest and brightest stars, thousands of times brighter and with diameters greater than 1000 times that of the Sun.

Gravity: Force of attraction between bodies, this depends on their mass and the distance between them. It holds the bodies of small mass in orbit of the larger bodies just as the Moon orbits the Earth and the Earth orbits the Sun

Hertzsprung-Russel diagram: Graph showing the relationship that shows the relationship between luminosities and spectral types of stars.

Horizon of events: The border of a black hole. After a body passes through the event horizon, it does not return anymore. As even the light escapes the gravitational force, we can not observe what takes place inside it.

Hubble constant: The number found dividing the speed of recession of a galaxy by the distance that separates it from the Earth. (This number is called the Hubble constant in memory of Edwin P. Hubble, the discoverer of the expanding universe.)

Light-years: Unit of distance equal to that traveled by the light in the vacuum in a year. One light-year equals 9.46 trillion kilometers, or 0.3066 Parsecs. A second light equals the distance light travels in a vacuum in one second and equals 299,792 kilometers.

Local group of galaxies: A cluster of at least 28 galaxies of which our own galaxy belongs.

Main sequence star: Star that lies within a well-defined diagonal band in the Hertzsprung-Russel diagram. Stars in the main sequence produce energy by the fusion of hydrogen to form helium at its nucleus.

Magnitude: Measure of the brightness of a star or other celestial body. Apparent magnitude is the brightness of an object as seen from Earth. Absolute magnitude is the magnitude an object would have if it were observed at a standard distance, which is 10 parsecs.

Meteor: Dust or rock particles that travel through space at high speed. A meteor (also called a shooting star) is the trail of light seen when a meteoroid burns as it fragments into the Earth's atmosphere. A meteorite is a larger meteoroid that enters the atmosphere and reaches the surface of the Earth. A meteor shower occurs when the Earth passes through a cloud of fragments in space.

Milky Way: The thin strip of light that runs through the night sky from the multiplicity of stars in our galaxy.

Moon: Natural satellite of a planet. This is also the name of Earth's only satellite.

Nebula: Gas cloud and interstellar dust. Nebulae are classified as emission nebulae, which shine; nebulae of reflection, which disperse the light of the stars. There are also dark nebulae that dim the starlight and more distant nebulae.

Neutrin: Elementary particle, with no electric charge and almost no mass, that moves with the speed of light. Neutrinos very rarely interact with any other matter.

Neutron star: Starburst that collapsed until it became almost entirely of neutrons. It has a mass between 1,5 and 3 solar masses, but has a very small diameter (usually about 10 km). Neutron stars are detected as pulsars.

Non - limiting condition: Theory that predicts a finite Universe, but without limits (in the imaginary time).

Open Universe: Theory or cosmological model in which the universe continues to expand forever.

Orbit: Curved trajectory of a body in space, influenced by the gravitational attraction of a body of greater mass. The orbital plane is the plane in which the orbit is described. The orbital slope is the angle between the orbital plane and a reference plane, for example the ecliptic. The orbital period is the interval of time that a body takes to complete an orbit.

Parsec: Unit distance equal 3.26 light years or 206 265 astronomical units.

Particle accelerator: An apparatus, such as a cyclotron or linear accelerator, that accelerates charged particles or nuclei, imparting them high speeds and high energies, useful in the research of subatomic particles

Perihelion: Point closest to Earth in the orbit of a planet or other body around the Sun.

Perigee: Point closest to Earth in the orbit of the Moon or an artificial satellite around Earth
 Phase: an apparent change in the form of a body, which depends directly on the light it receives from the star from which it orbits.
Planet: Relatively large body in orbit around the Sun or other star. The brightness of the planets is caused by the light reflected from the orbiting star.

Proto-Star: an initial stage in the life of a star, when it begins to condense into a Great Molecular Cloud, but can only be considered a proto-star until it enters the main sequence.
Proton: Particle with large mass and positive electric charge, found in the nucleus of the atoms; It consists of two quarks up and one quark down.

Press: regularly pulsating source of radio waves (sometimes of light and other radiations). The pulsars are believed to be rotating neutron stars.

Quasar: Compact object, extremely luminous, that appears with a star when seen from Earth. Little is known about quasars, and it is believed that they are likely active galactic nuclei, possessing a giant black hole as an energy source.

Radiation: Waves or particles emitted by some celestial body. The electromagnetic radiation is energy shifting in wave form, including gamma rays, x-rays, ultraviolet radiation, visible light, infrared radiation, microwave and radio waves. Particle radiation includes elementary particles, such as protons and electrons.

Retrograde movement: Movement contrary to the movement of the watch hands (when viewed from above).

Ring system: Fine disk of dust, rocks or ice particles orbiting the equatorial plane of some planets.

Satellite: Body orbiting a larger main body. The natural satellites of the planets are called moons. The artificial satellites have been placed in orbit of the Earth, the Moon and other planets.

Space-time curvature: According to Einstein's general theory of relativity, the effect caused in space by the presence of matter. (Gravity is seen as the consequence of the curvature of space induced by the presence of objects with large masses).

Shift to red: Displacement towards the longest long wavelengths of spectral lines of light from the stars of distant galaxies; occurs because these stars are moving away from the Earth.

Spectrum: A band or series of lines of electromagnetic radiation produced by radiation dispersion at its wavelengths, we can use as an example the colors of the rainbow: they are caused by the dispersion of white light that passes through the droplets of water that exist in the atmosphere .

Star: Astro luminous composed of gas, which shines due to the energy generated by nuclear reactions that occur at its core.

Singularity: A point in spacetime where its curvature becomes infinite: a term used by physicists and mathematicians to designate the point in the universe in which the equations of Einstein's general theory of relativity cease to exist; the moment of the Big Bang, when all the matter of the universe was contained in a single point.

Solar system: System comprising the Sun, the eight planets and other bodies orbiting around it due to gravity.

Sun: Central star of the Solar System. It is a star of the main sequence of medium size and density.

Supernova: A catastrophic blast of a massive star at the end of its life, during which it can become as bright as an entire galaxy. In the end, what remains of a supernova is an expanding cloud.

String theory: Theory according to which the elementary particles consist of tiny cords.

Theories of Great Unification (GUTS): Theories that try to prove that strong interactions, weak interactions and electromagnetic interactions are different aspects of a single fundamental force. (The ultimate goal is to incorporate gravitational interaction into these same all-pervasive theories).

Universe in expansion: The idea, first proposed by the American astronomer Edwin Hubble in 1929, that galaxies are moving away from Earth, and one another, at a constant rate. Closed Universe: The cosmological theory that conceives of the expanding universe as "closed" or destined to stop expanding at some future time, which would be followed by the collapse of all galaxies into a backward-looking Big Bang species, new phase of expansion.

Variable star: Stars whose brightness varies.

Wave-length: In a wave, the distance between two continuous intervals (or two crests).

White dwarf: A collapsed star, small and very dense, which cools gradually, supported by the repulsion of the principle of electron exclusion.

Year: It is the time interval that a planet takes to complete an orbit around the Sun. The sidereal year is the time interval spent in describing a measured orbit using the fixed stars as a positional reference point. A tropic year is the interval of time that is measured using a specific position of the sun under the celestial sphere.

References

A brief history of time: from the Big Bang to black holes. Stephen H. 1988. Space time publications.

Black holes and time Warps. Kip Thorne. WW Norton. 1994.

Black holes physics in an electromagnetic. Wave Guide. Steve K. Blau, Physics today. Vol. 58, no. 8, pp. 19-20; August 2005.

Dark energy. Robert R. Caldwell. physics World. Vol. 17, no. 5, pp. 37-42; May 2004.

The elegant universe - Super strings, hidden dimensions and the search for definitive theory. Brian Greene, company of letters; 2001.

Planetesimals to brown dwarfs: what is a planet? Gibor Bostli and Michael E. Brawn. Annual Review of Earth and Planetary Sciences. Vol. 34, p. 193-216; 2006.

Supernova explosions in the Universe. A. Burrons. Nature. Vol. 403, p. 727-733; February 17, 2000.

The five ags of the Universe: inside the physics of eternity. Fred C. Adams and Greg Laughlin, Free Press; 2000.

The infiniti Cosmos: questions from the frontiers of cosmology. Joseph Silk. Oxford University Press; 2006.

The "misterious" origin of brown dwarfs. Paolo Padoan and Åke Nordlund. Astrophysical Journal, vol. 626 No. 1, part 1, pp. 498-522; July 10, 2005.

The return of a Static Universe and the end of cosmology. Lawrence M. Kraus and Robert J. Scherrer. Journal of General Relativity and Gravitation. Vol. 39, no. 10, p. 1545-1550; October 2007.

The T Tauri phase down to nearly planetary masses. Subhanjoy, Ray Jayawardhana and Gibor Basri. Astrophysical Journal. Vol. 626, no. 1, part 1, p. 498-522; July 10, 2005.

Towards a deterministic model of planetary formation. S. Ida and DNC Lin. Astrophysical Journal, Vol. 604, No. 1, pp. 388-413. March 2004.

What is a planet? Steven Soter. Astronomical Journal. Vol. 132, No. 6, pp. 2513-2519. December 2006.